October 26, 2017

Dear Customer:

Thank you for your p[illegible] of *Evol*[illegible]*ary Multi-Objective System Design: Theory and Applications*, by Nadia Nedjah, Luiza De Macedo Mourelle, and Heitor Silverio Lopes.

On page xi, in the Table of Contents, Mouloud Koudil is incorrectly listed as a contributor to Chapter 9,

On page 171, the first page of Chapter 9, Mouloud Koudil is incorrectly listed as a chapter contributor.

We sincerely regret any inconvenience this may have caused you. Please
let us know if we can be of any assistance regarding this title or any other titles that Taylor & Francis publishes.

Best regards,
Taylor & Francis
978-1-4987-8028-5

Evolutionary Multi-Objective System Design

Theory and Applications

CHAPMAN & HALL/CRC
COMPUTER and INFORMATION SCIENCE SERIES

Series Editor: Sartaj Sahni

PUBLISHED TITLES

ADVERSARIAL REASONING: COMPUTATIONAL APPROACHES TO READING THE OPPONENT'S MIND
Alexander Kott and William M. McEneaney

COMPUTER-AIDED GRAPHING AND SIMULATION TOOLS FOR AUTOCAD USERS
P. A. Simionescu

DELAUNAY MESH GENERATION
Siu-Wing Cheng, Tamal Krishna Dey, and Jonathan Richard Shewchuk

DISTRIBUTED SENSOR NETWORKS, SECOND EDITION
S. Sitharama Iyengar and Richard R. Brooks

DISTRIBUTED SYSTEMS: AN ALGORITHMIC APPROACH, SECOND EDITION
Sukumar Ghosh

ENERGY-AWARE MEMORY MANAGEMENT FOR EMBEDDED MULTIMEDIA SYSTEMS: A COMPUTER-AIDED DESIGN APPROACH
Florin Balasa and Dhiraj K. Pradhan

ENERGY EFFICIENT HARDWARE-SOFTWARE CO-SYNTHESIS USING RECONFIGURABLE HARDWARE
Jingzhao Ou and Viktor K. Prasanna

EVOLUTIONARY MULTI-OBJECTIVE SYSTEM DESIGN: THEORY AND APPLICATIONS
Nadia Nedjah, Luiza De Macedo Mourelle, and Heitor Silverio Lopes

FROM ACTION SYSTEMS TO DISTRIBUTED SYSTEMS: THE REFINEMENT APPROACH
Luigia Petre and Emil Sekerinski

FROM INTERNET OF THINGS TO SMART CITIES: ENABLING TECHNOLOGIES
Hongjian Sun, Chao Wang, and Bashar I. Ahmad

FUNDAMENTALS OF NATURAL COMPUTING: BASIC CONCEPTS, ALGORITHMS, AND APPLICATIONS
Leandro Nunes de Castro

HANDBOOK OF ALGORITHMS FOR WIRELESS NETWORKING AND MOBILE COMPUTING
Azzedine Boukerche

HANDBOOK OF APPROXIMATION ALGORITHMS AND METAHEURISTICS
Teofilo F. Gonzalez

HANDBOOK OF BIOINSPIRED ALGORITHMS AND APPLICATIONS
Stephan Olariu and Albert Y. Zomaya

PUBLISHED TITLES CONTINUED

HANDBOOK OF COMPUTATIONAL MOLECULAR BIOLOGY
Srinivas Aluru

HANDBOOK OF DATA STRUCTURES AND APPLICATIONS
Dinesh P. Mehta and Sartaj Sahni

HANDBOOK OF DYNAMIC SYSTEM MODELING
Paul A. Fishwick

HANDBOOK OF ENERGY-AWARE AND GREEN COMPUTING
Ishfaq Ahmad and Sanjay Ranka

HANDBOOK OF GRAPH THEORY, COMBINATORIAL OPTIMIZATION, AND ALGORITHMS
Krishnaiyan "KT" Thulasiraman, Subramanian Arumugam, Andreas Brandstädt, and Takao Nishizeki

HANDBOOK OF PARALLEL COMPUTING: MODELS, ALGORITHMS AND APPLICATIONS
Sanguthevar Rajasekaran and John Reif

HANDBOOK OF REAL-TIME AND EMBEDDED SYSTEMS
Insup Lee, Joseph Y-T. Leung, and Sang H. Son

HANDBOOK OF SCHEDULING: ALGORITHMS, MODELS, AND PERFORMANCE ANALYSIS
Joseph Y.-T. Leung

HIGH PERFORMANCE COMPUTING IN REMOTE SENSING
Antonio J. Plaza and Chein-I Chang

HUMAN ACTIVITY RECOGNITION: USING WEARABLE SENSORS AND SMARTPHONES
Miguel A. Labrador and Oscar D. Lara Yejas

IMPROVING THE PERFORMANCE OF WIRELESS LANs: A PRACTICAL GUIDE
Nurul Sarkar

INTEGRATION OF SERVICES INTO WORKFLOW APPLICATIONS
Paweł Czarnul

INTRODUCTION TO NETWORK SECURITY
Douglas Jacobson

LOCATION-BASED INFORMATION SYSTEMS: DEVELOPING REAL-TIME TRACKING APPLICATIONS
Miguel A. Labrador, Alfredo J. Pérez, and Pedro M. Wightman

METHODS IN ALGORITHMIC ANALYSIS
Vladimir A. Dobrushkin

MULTICORE COMPUTING: ALGORITHMS, ARCHITECTURES, AND APPLICATIONS
Sanguthevar Rajasekaran, Lance Fiondella, Mohamed Ahmed, and Reda A. Ammar

PERFORMANCE ANALYSIS OF QUEUING AND COMPUTER NETWORKS
G. R. Dattatreya

PUBLISHED TITLES CONTINUED

THE PRACTICAL HANDBOOK OF INTERNET COMPUTING
Munindar P. Singh

SCALABLE AND SECURE INTERNET SERVICES AND ARCHITECTURE
Cheng-Zhong Xu

SOFTWARE APPLICATION DEVELOPMENT: A VISUAL C++®, MFC, AND STL TUTORIAL
Bud Fox, Zhang Wenzu, and Tan May Ling

SPECULATIVE EXECUTION IN HIGH PERFORMANCE COMPUTER ARCHITECTURES
David Kaeli and Pen-Chung Yew

TRUSTWORTHY CYBER-PHYSICAL SYSTEMS ENGINEERING
Alexander Romanovsky and Fuyuki Ishikawa

VEHICULAR NETWORKS: FROM THEORY TO PRACTICE
Stephan Olariu and Michele C. Weigle

X-Machines for Agent-Based Modeling: FLAME Perspectives
Mariam Kiran

Evolutionary Multi-Objective System Design

Theory and Applications

Edited by

Nadia Nedjah

State University of Rio de Janeiro, Brazil

Luiza De Macedo Mourelle

State University of Rio de Janeiro, Brazil

Heitor Silverio Lopes

Federal University of Parana, Brazil

CRC Press is an imprint of the
Taylor & Francis Group, an **informa** business

A CHAPMAN & HALL BOOK

CRC Press
Taylor & Francis Group
6000 Broken Sound Parkway NW, Suite 300
Boca Raton, FL 33487-2742

Printed on acid-free paper
Version Date: 20170826

International Standard Book Number-13: 978-1-4987-8028-5 (Hardback)

Visit the Taylor & Francis Web site at
http://www.taylorandfrancis.com

and the CRC Press Web site at
http://www.crcpress.com

Contents

Preface

Real-world engineering problems often require concurrent optimization of several design objectives, which are conflicting in most of the cases. Such an optimization is generally called multi-objective or multi-criterion optimization. The area of research that applies evolutionary methodologies to multi-objective optimization is of special and growing interest. It brings a viable computational solution to many yet-opened real-world problems.

Generally, multi-objective engineering problems do not have a straightforward optimal design. These kind of problems usually inspire several solutions of equal efficiency, which achieve different trade-offs of the considered objectives. Decision maker's preferences are normally used to select the most adequate design. Such preferences may be dictated before or after the optimization takes place. They may also be introduced interactively at different levels of the optimization process. Multi-objective optimization methods can be subdivided into classical and evolutionary. The classical methods usually aim at a single solution while the evolutionary methods provide a whole set of so-called Pareto-optimal solutions.

This book provides a representation of the state of the art of evolutionary multi-objective optimization research areas and related new trends. Especially, it reports many innovative designs yielded by the application of such optimization methods. In the following, we give a brief description of the main contribution of each of the included chapters.

In the first chapter, entitled "Embrittlement of Stainless Steel Coated Electrodes," the authors aim at creating a mathematical model using Artificial Neural Networks, which represents the manufacturing process of the stainless steel coated electrodes. The proposed model allows to find the most relevant input variables, from the statistical point of view, reducing the cost of experimentation. Another achievement of this work is to apply multi-objective optimization techniques to achieve the best process settings, seeking to reduce the residual moisture and the fragility of the product as well as to maintain or to reduce the manufacturing costs.

In the second chapter, entitled "Learning Fuzzy Rules from Imbalanced Datasets using Multi-objective Evolutionary Algorithms," the authors describe a method to learn fuzzy classification rules from imbalanced datasets using multi-objective genetic algorithms and the iterative rule learning approach. In this approach, a single rule is learned in each execution of the multi-objective evolutionary algorithm. The proposed method contains two phases: preprocessing, to balance the imbalanced dataset; and iterative fuzzy rule learning, to learn the fuzzy rule base. The performance of the method under the preprocessing algorithms is analyzed and discussed. The results show that our method achieved better performance in most of the 24 datasets used in the experiments regarding accuracy, number of rules, and number of conditions when compared to a similar method found in the literature.

In the third chapter, entitled "Hybrid Multi-Objective Evolutionary Algorithms with Collective Intelligence," the authors propose two new collective preference-based interactive Multi-objective Evolutionary Algorithms based on the non-dominated sorting genetic algorithm II and the S-metric selection algorithm, respectively. These algorithms combine two distinct fields not connected before as it aggregates consistent preferences from a collective intelligence environment to the evolutionary optimization process. These use people's heterogeneity and common sense to support one or many experienced decision makers in the search for relevant regions in Pareto-optimal set. Built upon the subjectivity of the crowds and human cognition, the intelligence of participatory actions addresses dynamic collective reference points to overcome the difficulties of multi-objective problems and guide the exploration of preferred solutions. The proposed method improves the quality of the obtained Pareto frontier approximation.

In the fourth chapter, entitled "Multi-objective Particle Swarm Optimization Fuzzy Gain Scheduling Control," the authors propose a fuzzy gain-scheduling controller design based on robust stability criterion via Multi-objective Particle Swarm Optimization for Takagi–Sugeno fuzzy model. The plant to be controlled is identified from input–output experimental data, by using the fuzzy C-Means clustering algorithm and least-squares estimator for antecedent and consequent parameters estimation, respectively. The MOPSO algorithm is used to tune the fuzzy gain scheduling controller parameters, via Parallel Distributed Compensation strategy, based on gain and phase margins specifications, according to identified fuzzy model parameters of the plant to be controlled for each operating condition. There are two objectives to be optimized, namely: obtaining the fuzzy gain scheduling controller parameters so as to guarantee the gain margin so near as possible of the specified gain margin; and obtaining fuzzy gain scheduling controller parameters so as to guarantee the phase margin so near as possible of the specified phase margin. These two objectives are conflicting. Experimental results for fuzzy gain scheduling control of a thermal plant with time varying delay are presented to illustrate the efficiency and applicability of the proposed methodology.

In the fifth chapter, entitled "Multi-objective Evolutionary Algorithms for Smart Placement of Roadside Units in Vehicular Networks," the authors focus on the efficient design of the roadside unit infrastructure, also known as the roadside unit deployment problem, which consists in selecting the best locations and roadside unit types in order to optimize both the service provided by the fixed infrastructure and the economical deployment costs. This is a tractable problem when dealing with small sized areas, but it results in a hard-to-solve problem for city-scaled instances, as the number of possible solutions becomes very large. Three Multi-Objective Evolutionary Algorithms are studied, applied, and their performance compared in solving an interesting case of study. There algorithms are a linear aggregation approach and the NSGA-II and SPEA2. One of the considered objectives consists of minimizing the economical cost of the deployment. The other objective consists of maximizing the network service provided by the platform. In this study, a specific QoS model is proposed, considering the number of vehicles, speed, and coverage of street segments in the city, and a Monte Carlo simulation approach is used to compute the corresponding QoS metric.

In the sixth chapter, entitled "Solving Multi-Objective Problems with MOEA/D and Quasi-Simplex Local Search," the authors describe an empirical analysis of the influence of the components and parameters of the local searches in the performance of the Multi-Objective Evolutionary Algorithm based on Decomposition. The best version is compared to the original version of the algorithm and to NSGA-II algorithms, which are two of the most cited and influential Multi-objective Evolutionary Algorithms in the literature. Four benchmarks commonly employed in the Multi-Objective Optimization literature (CEC 2009, WFG, DTLZ, and ZDT) are used to evaluate the algorithms. The quality of the Pareto approximations is measured in terms of three distinct quality indicators: hypervolume, addictive unary-ϵ, and IGD.

In the seventh chapter, entitled "Multi-objective Evolutionary Design of Robust Substitution Boxes," the authors focus on a two-fold objective: first, they evolve a regular S-box with high non-linearity and low auto-correlation properties using evolutionary computation; then they automatically generate evolvable hardware for the obtained S-box. Targeting the former, they use the Nash equilibrium-based multi-objective evolutionary algorithm to optimize regularity, non-linearity, and auto-correlation, which constitute the three main desired properties in resilient S-boxes. Pursuing the latter, they exploit genetic programming to automatically generate the evolvable hardware designs of substitution boxes that minimize hardware space, encryption/decryption time, and dissipated power, which form the three main hardware characteristics. They compare the obtained results against existing and well-known designs, which were produced using conventional methods as well as through evolution.

In the eighth chapter, entitled "Multi-objective Approach to the Protein Structure Prediction Problem," the authors propose the application and comparison of a multi-objective approach to the Protein Folding Problem, considering two objectives. The main objective is to minimize the energy calculated from the Hydrophobic–Hydrophilic model, and the second objective consists of minimizing the Euclidean distance between amino acids of the protein. The introduction of the second objective is inspired by existing work, in which the evaluation of a structure represented by this model considers only the number of hydrophobic contacts, disabling the optimization algorithms to distinguish between structures with the same number of hydrophobic contacts. Using a multi-objective approach, other characteristics of the protein, including its energy, are investigated. In particular, the authors investigate the distance because more compact structures tend to have more hydrophobic contacts due to the fact that the lower the Euclidean distance between the amino acids is, the more compact the whole conformation will be. For this purpose, the authors exploited two versions of each of the NSGA-II and IBEA algorithms. A backtrack strategy is used to generate the initial population to avoid the generation of many invalid solutions.

In the ninth chapter, entitled "Multi-objective IP Assignment for Efficient NoC-based System Design," the authors propose a multi-objective evolutionary based decision system to help Network-on-Chip designers to select the most IP blocks suitable used during the assignment step. To this aim, they use the structural representation of task graphs and intellectual properties repository data from the Embedded Systems Synthesis benchmarks Suite (E3S) as an IP library for the proposed tool. They exploit the multi-objective particle swarm optimization algorithm, which was modified to suit the specificities of the assignment problem and also to guarantee the NoC designer's constraints.

The editors of this book are very much grateful to the authors for their valuable contributions and the reviewers for their tremendous service by critically reviewing the submitted works. The editors would also like to thank the editorial team that helped to format this work in a nice book. Finally, we sincerely hope that the reader will share our excitement with this book on reconfigurable and adaptive computing and will find it useful.

Nadia Nedjah
State University of Rio de Janeiro, Brazil
Luiza M. Mourelle
State University of Rio de Janeiro, Brazil
Heitor Silvério Lopes
Technological Federal University of Paraná, Brazil

List of Figures

List of Tables

Chapter 1

Embrittlement of Stainless Steel Coated Electrodes

Diego Henrique A. Nascimento

Graduate Program in Mathematical and Computational Modelling, Centro Federal de Educação Tecnológica de Minas Gerais, Belo Horizonte, MG, Brazil

Rogério Martins Gomes

Intelligent Systems Laboratory, Centro Federal de Educação Tecnológica de Minas Gerais, Belo Horizonte, MG, Brazil

Elizabeth Fialho Wanner

Department of Computer Engineering, Centro Federal de Educação Tecnológica de Minas Gerais, Belo Horizonte, MG, Brazil

Mariana Presoti

Graduate Program in Mathematical and Computational Modelling, Centro Federal de Educação Tecnológica de Minas Gerais, Belo Horizonte, MG, Brazil

---

1.1 Introduction

Union of structures or metallic parts in the metalworking industry is usually accomplished by electric arc welding [163] [38]. Coated electrodes are welding consumables composed of two elements, the electrode core wire and the ceramic coating. The welding process is performed when the coated electrode is attached to the welding equipment, which is energized with a polarity, while the metals to be joined are energized with reverse polarity [139]. When the electrode comes into contact with the workpiece, the electrode is melted by the electric arc formed in this process, leading the union of the parts or joint.

There are various types of coated electrodes, each of which is applied for a specific application. Thus, in order to guarantee the adequacy of the electrode to the type of application the American Welding Society standard (AWS)[7] is used. This standard is responsible for regulating the classification of each type of electrode and for establishing the chemical composition and quality requirements of the consumables [135].

In the manufacturing process, the electrode must go through many stages of drying in order to achieve the moisture level established by AWS. If the heat input is undersized (low temperature and low exposure time), the electrode will not achieve the moisture levels required [44]. On the other hand, if the heat input is oversized, there will be a high evaporation rate in the coating, leading to cracks or coating adhesion problems to the electrode core wire caused by the effect of differential expansion. This differential expansion occurs due to the fact that the electrode core wire and its coating do not present the same level of expansion when heated nor the same elasticity [34]. At the beginning of the drying process, the electrode core wire and the coating dilate at the same speed. However, at the moment of cooling, the electrode core wire contracts more rapidly than the coating, causing gaps between them. Consequently, the ceramic coating becomes more vulnerable to mechanical impact [34] arising from the process of packing and shipping. These impacts fracture the coating that takes off from the metal core, rendering an inoperable product. Therefore, a suitable heat input for each electrode must be well understood to avoid losses and rework during manufacturing process. This process, called embrittlement of the electrode, invalidates the process of packing and shipping of the product by coating breaks.

Taking into account the coated electrodes, there is a special family of products called stainless steel electrodes that has a high rate of rejection during the production process due to the high level of fragility. Rejections that occur due to the fragility of the product entail a loss of US$ 6,798.00 per ton in the company used as a model in this work. Thus, the high cost of manufacturing combined with the high cost of rejection makes it important to develop a methodology that attempts to mitigate or eliminate this sort of problem.

Bearing this in mind and in order to understand the relations present in each phase of this process, this work aims at creating a mathematical model using Artificial Neural Networks (ANN) which represents the manufacturing process (Figure 1.1) of the stainless steel coated electrodes. This mathematical model allows to find the most relevant input variables, from the statistical point of view, reducing the cost of experimentation. Another achievement of this work is to apply multi-objective optimization techniques to achieve the

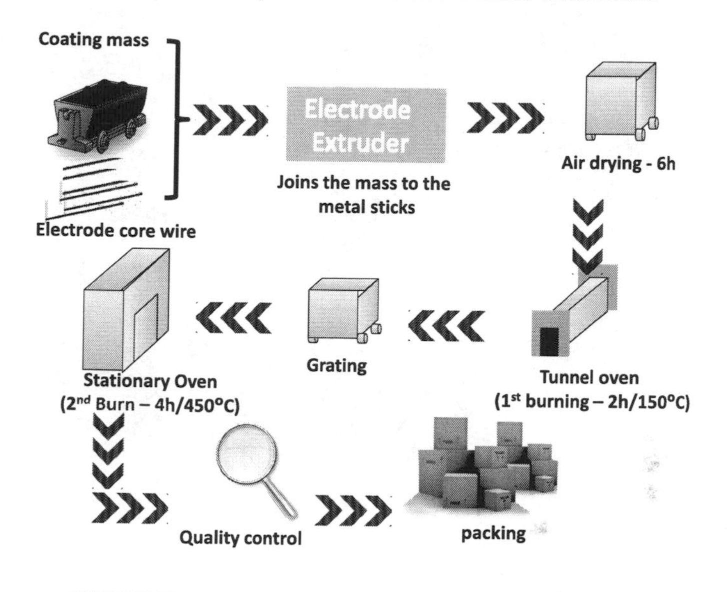

FIGURE 1.1: Flowchart of manufacturing process.

best process settings, seeking to reduce the residual moisture and the fragility of the product as well as to maintain or to reduce the manufacturing costs.

For this purpose, we used two mathematical models that establish the relation between the process parameters and the output variables (fragility, moisture, and cost). The first model was created using a Multi-Layer Artificial Neural Network (MLP) [119]. This model represents the manufacturing process and is capable of providing, based on the process input parameters, an output related to the moisture index and fragility of the product (Figure 1.2). The second model, in turn, is a empirical function of the manufacturing average cost based on the process parameters. These models will compose the objective functions of the computational optimization process.

1.2 Manufacturing Process

The coated electrodes' manufacturing process can be represented by a chain of complex steps, as shown in Figure 1.1. However, in this work, only the electrode drying process will be studied. Its stages can be summarized as follows:

1. **Air-drying**: after the manufacturing process, the electrodes are sent to the standby courtyard, where they remain for 6 hours. During this step, the product loses much of the moisture and acquires some rigidity to withstand the other drying stages.

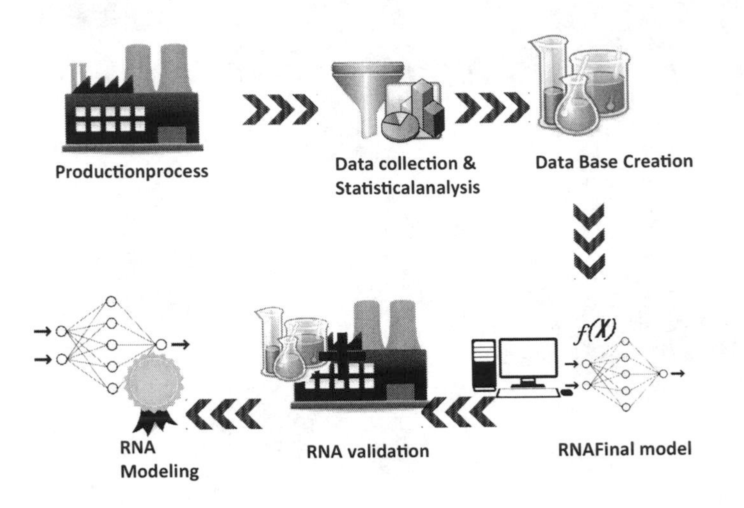

FIGURE 1.2: Flowchart of the modeling process.

2. **First Burning**: after the air-drying, the electrodes are forwarded to a tunnel oven where the product is drawn through by pneumatic actuators. The electrodes remain in this crossing process for 2 hours. This oven is maintained at a temperature of 150 °C and, after crossing, the electrodes are removed and allocated to a standby courtyard, waiting to be sent to the second burning stage.

3. **Second Burning**: after the first burning, the electrodes are forwarded to a stationary oven where they remain for 4 to 5 hours at a temperature of 450 °C. After this process, the electrodes are removed and sent to the packaging step and quality control, ending the production process.

The following variables are present in each stage of the manufacturing process:

- **Ambient Relative Humidity and Ambient Temperature**: This value depends on the season, year, time of the day, and measuring point.

- **Exposure Time**: process parameter that determines the air-drying and heat exchange time at each stage of the process. The manufacturing process takes 6 hours in the air-drying stage, 2 hours in the first burning, and 4 to 5 hours in the second burning.

- **Oven Temperature**: process parameter that determines the exposure temperature of the ovens during the drying process. This parameter is present in the first and second burnings at 150 °C and 450 °C, respectively.

These input variables are relevant to define and to produce a product that meets the quality established by the AWS standard. The ambient relative humidity and ambient temperature are measured through thermo-hygrometers installed at different points of the process. The temperatures of the oven, in turn, are controlled by a control panel, where it is possible to select a specific temperature as well as the exposure time that the product will remain in each stage.

The output variables of the manufacturing process are dependent on the input variables and can be described as follows:

- **Fragility**: electrode coating resistance to absorb impacts. The objective value at the end of the process is to keep the fragility level below 3%. This value was established empirically, based on the manufacturing history of the model company used in this study. According to this company, this limit of fragility ensures the transport and the safe handling of the product.

- **Product Moisture**: residual amounts of water present in the coating after the drying process. The AWS standard [7][138] establishes that the moisture level of the product must be kept lower than 0.50%.

The measurement of the coating moisture is perfomed by the equipment RC412 manufactured by LECO, which measures the release rate of H_2O of a sample. This device heats the coating at 600 °C and measures the amount of molecules of water released from the coating structure [172].

The measurement of the fragility is performed by the drop test, which simulates the impact which occurs in the packaging process and vibration concerning the transport of the product. This test is performed using a constant height platform related to the diameter of the product. The electrode is released from this platform over another metal platform. After the fall, the mass of the electrode is measured (m_{after}) and compared to the mass before the fall (m_{before}). At the end, the influence of the mass (m_{wire}) must be discounted in order to obtain the final result of fragility. Equation 1.1 calculates the fragility of the product:

$$F(m_{before}, m_{wire}, m_{after}) = \frac{m_{before} - m_{after}}{m_{before} - m_{wire}}. \qquad (1.1)$$

As defined above, the acceptance criteria for final approval of the product are fragility and moisture values lower than 3% and 0.50%, respectively.

1.3 Process Modeling

The study of production processes usually is based on real experiments that involve high costs. These costs are related to factory resources and to the minimum amount that must be produced in order to ensure that the products are in compliance with technical specifications. The cost involved for performing a production test of the stainless steel electrodes may reach US$1,359.60. Thus, in order to minimize the trial costs, this work proposes to develop a model of tests in a laboratory which represents a real manufacturing process. Laboratory tests may be conducted under a product sampling of 500 g, maximum load capacity of the equipment used in the laboratory, reducing the cost to US$3.55 per test.

In order to carry out these tests some considerations should be established. Usually, the ambient temperature and relative humidity are kept constant in laboratory tests. However, in field tests, these variables can not be controlled. Therefore, the first step is to certify if changes in ambient temperature and humidity, which occur naturally during air-drying stage, are statistically significant.

The variables involved in the manufacturing process were monitored in the air-drying stage. During the process, five groups of measurements were performed concerning the input variables, ambient temperature, and relative humidity, in the first week of each month, beginning in March, during five alternate months. In the air-drying stage, five measurements of the ambient temperature and relative humidity were carried out throughout the day, starting at 6:00 a.m., for each measurement group. The total exposure time for the product in this stage was 6 hours. Samples were collected using a digital thermo-hygrometer and can be seen in Tables 1.1 and 1.2.

TABLE 1.1: Ambient temperature (°C)

	6h	7h	9h	11h	12h
Week 1	29.4	29.4	28.9	30.3	30.3
Week 2	31	31	29.5	29.1	28.1
Week 3	30	30	28.5	26.7	35.7
Week 4	29.7	29.7	27.6	24.7	35.7
Week 5	31	31	32.8	33.7	36.7

TABLE 1.2: Ambient relative humidity (%)

	6h	7h	9h	11h	12h
Week 1	42	42	43	38	35
Week 2	40	40	42	38	40
Week 3	40	40	46	47	30
Week 4	48	40	47	46	38
Week 5	46	48	36	30	35

In order to verify that natural variations in temperature and humidity are statistically relevant, a technique of analysis of variance, ANOVA [179], was applied. ANOVA is a statistical tool to assess indicators of a population, such as mean and variance, providing information regarding the significance of them. However, some statistical requirements must be observed to apply the ANOVA, i.e., normality, randomness, and homoscedasticity of the data. Tests were carried out on the data (Tables 1.1 and 1.2) and confirmed the aforementioned conditions for applying ANOVA. Thus, following the ANOVA method, it is necessary to define a null hypothesis.

The null hypothesis to be considered in this work establishes that the difference between the average of the ambient temperature and humidity in respect to each one of the measurements tends to zero, indicating that the average is a good indicator for this group of measurements. If this assumption is accepted, i.e., it is not distorted, the variation of the ambient temperature and humidity during all five measurements is not statistically significant. Tables 1.3 and 1.4 show the results of ANOVA regarding ambient temperature and humidity, respectively. Once the values of $F0$ do not exceed the standard values (3.13), one can not reject the null hypothesis, indicating that there is not statistical evidence to say that the values of ambient temperature and humidity are different along the measurements. Thus, these input variables can be kept constant within the laboratory environment and, therefore, the exposure time in the air will be the only input variable to be considered in the air-drying stage.

TABLE 1.3: ANOVA–Ambient temperature

	S.Q	GL	MQ	F0
Time	0.02	4	0.01	3.08
Error	0.07	40	0.00	
Total	0.09	44		

TABLE 1.4: Anova–Ambient relative humidity

	S.Q	GL	MQ	F0
Time	54.80	4	13.70	2.95
Error	185.78	40	4.64	
Total	240.58	44		

Another consideration to be made regarding the laboratory test is whether the results of the output variables, moisture, and fragility of the product follow a normal distribution during the stages of first and second burnings. In order to answer this question, a new experiment was conducted. During the stages of first and second burnings, the temperatures and exposure times were previously set at [150 °C, 450 °C] and [2 h, 4 h], respectively. These values were automatically controlled by the ovens and were checked twice (at the beginning and at the end of the burning stages) in all five weeks of measurement. The evolution of the output variables, fragility, and moisture of the product were assessed on the data collected. Tables 1.5 and 1.6 show the measured values for moisture and fragility of the product during the burning stages. The hypothesis of normality, randomness, and homoscedasticity of the measured data was observed for each of the output variables. Thus, it is possible to conclude that the results of the output variables during the two burning stages follow a normal distribution.

TABLE 1.5: Moisture of the product in the ovens

	1st Burn - Q_1		2nd Burn - Q_2	
	before	after	before	after
Week 1	1.5%	0.8%	0.8%	0.22%
Week 2	1.7%	0.7%	0.7%	0.12%
Week 3	1.8%	0.7%	0.7%	0.20%
Week 4	1.9%	0.8%	0.8%	0.18%
Week 5	1.5%	0.8%	0.8%	0.21%

TABLE 1.6: Fragility of the product in the ovens

	1st Burn - Q_1		2nd Burn - Q_2	
	before	after	before	after
Week 1	0.00%	0.44%	0.4%	0.6%
Week 2	0.05%	0.40%	0.4%	3.8%
Week 3	0.10%	0.90%	0.9%	5.3%
Week 4	0.05%	0.40%	0.4%	0.6%
Week 5	0.03%	0.70%	0.7%	4.4%

After the statistical tests and data normality verification one can conclude that the manufacturing process can be modeled using techniques based on normal distribution. Furthermore, the process of air-drying can be simulated using temperature and humidity constants, since they have no influence on the output variables in the laboratory tests.

1.3.1 ANN Database

After the preliminary measurements and analysis of the parameters, a model of the production process based on ANN was created. The process modeled by ANN comprises the stages of air-drying, first, and second burnings; input variables: exposure time in each stage and temperature of each burning; output variables: moisture and fragility of the product.

The model was developed using two types of equipment, a ventilated oven and an electric muffle oven. The ventilated oven simulates the air-drying and the first burning stages. The muffle, in turn, simulates the second burning. For carrying out the tests, electrodes of the same production batch were collected at the factory and taken to the laboratory to perform the controlled drying process.

Using the aforementioned equipment, a series of tests were designed to raise a database containing several input patterns and their respective outputs, in regards to the manufacturing process. The input values were defined empirically by observing a range of variation for each input. The air-drying time variable has a range of 6 hours divided into intervals of 1 hour and the temperature variables of first and second burnings are limited to 200 °C and 500 °C, respectively. Therefore, tests were defined to temperatures of 100 °C, 150 °C, and 200 °C for the first burning and 400 °C, 450 °C, and 500 °C for the second burning. These variations were chosen arbitrarily to compose an initial database of 144 tests.

1.3.2 MultiLayer Perceptron Network — MLP

The MLP network was implemented on 2 platforms, MATLAB® and JustNN. MATLAB is a platform that provides mathematical implementations through a high level programming language and offers a wide range of libraries, facilitating the construction of complex architectures [109]. JustNN is a free software made by the company Neural Planner Software (http://www.justnn.com/) and aims to simplify the implementation of ANN.

The database generated in the laboratory was divided randomly into two parts, 80% for training and 20% for validation. Several tests were performed by varying the number of layers, number of neurons per layer, and training algorithms. The training steps were repeated five times and the performance indicators, including the regression coefficient during training and accuracy rate, were calculated using the average of runs. After testing, the Levenberg–Marquardt algorithm was chosen for presenting the best performance for solving this particular problem. Table 1.7 summarizes the network architectures that presented the best results.

TABLE 1.7: Artificial neural network tests

	N° hiden layers	Act. function	MSE error (meta)	Epochs
Just NN	6x4x4	Purelin	0.10%	100
MATLAB	6x4x4	Purelin	0.10%	100

After training it was observed that the accuracy rate for the network implemented using the JustNN platform was 93% while the accuracy rate for the MATLAB platform was 71%. Furthermore, the JustNN platform also presented higher regression coefficients and lower error rate. This rate was measured considering an acceptance of the result within a range of +/− 10% compared to the original data. Based on the best results, the network chosen to model the manufacturing process was implemented in JustNN platform.

Finally, one last test was proposed, in which three new standard processes were chosen to be tested in a real manufacturing process in order to validate the adequacy of the model. The results for these new settings are shown in Table 1.8. The error obtained between the real case and the simulated model was less than 20% but despite this magnitude, it was possible to observe that in all cases, the results were satisfactory provided that the fragility rate was less than 3%.

TABLE 1.8: Modeling performance certification

	Moisture		Brittleness	
	Simulated	Real	Simulated	Real
Process 1	0.29%	0.29%	2.77%	2.84%
Process 2	0.29%	0.36%	2.8%	2.6%
Process 3	0.29%	0.31%	2.71%	2.52%

1.4 Process Optimization

Bearing in mind the quality optimization process of coated electrodes, it is relevant to observe the binomial, production cost versus product quality. Often these characteristics are antagonistic, i.e., there is no acceptable solution that guarantees the best value in all aspects of evaluation. In this case, an increase in the product quality usually implies an increase in the production cost, which could impair the competitiveness in the market. Thus, a manufacturing cost function to be used in combination with the quality function was proposed. In this cost function the following variables are considered:

- **Monthly Maintenance Cost** (M): preventive and corrective maintenance costs for each oven, first and second burnings, per hour.
- **Operational Cost** (O): labor costs per hour worked.
- **Energy Cost** (E): average cost of electricity or natural gas used in the drying process, per hour.
- **Time of the Production Cycle** (t): time duration of a complete production cycle.
- **Process Capability**(Z): productive capacity of each stage of the process measured in tons per hour.
- **Temperature of the Ovens** (T): working temperature in each oven.

The temperature of the ovens act as a weight of energy costs in the drying stages. Considering that the present method uses a temperature of 150 °C for first burning and 450 °C for second burning, temperature values will be standardized based on these values. Thus, temperatures above or below this current value represent an increase or a decrease in energy costs, respectively.

Equation 1.2 represents the production cost per ton concerning the air-drying stage.

$$Cost_{air} = \frac{t_{air}.O_{air}}{Z_{air}}. \tag{1.2}$$

Equations 1.3 and 1.4 represent the cost relating to the first and second burning stages.

$$\begin{aligned} Cost_{1Q} = & \frac{t_{1B}.O_{1B}}{Z_{1B}} + \frac{M_{1B}.t_{1B}}{Z_{1B}} + \\ & \frac{E_{\text{Gas}}.t_{1B}.T_{1B}}{150.Z_{1B}}. \end{aligned} \tag{1.3}$$

$$\begin{aligned} Cost_{2B} = & \frac{t_{2B}.O_{2B}}{Z_{2B}} + \frac{M_{2B}.t_{2B}}{Z_{2B}} + \\ & \frac{E_{\text{electrical}}.t_{2B}.T_{2B}}{450.Z_{2B}}. \end{aligned} \tag{1.4}$$

Equations 1.2, 1.3, and 1.4 were added in Equation 1.5 in order to obtain the total production cost per ton, considering all process parameters.

$$F_{cost} = Cost_{air} + Cost_{1B} + Cost_{2B} \tag{1.5}$$

1.4.1 Multi-objective Optimization

The multi-objective optimization problem can be formulated as follows:

$$\begin{aligned} & x^* = \min_x f_k(x); k = 1, 2, ..., m \\ & \text{subject to:} \begin{cases} g_i(x) \leq 0; & i = 1, 2, \cdots, r \\ h_j(x) = 0; & j = 1, 2, \cdots, p \end{cases}. \end{aligned} \tag{1.6}$$

where $x \in \mathbb{R}^n$, $f(\cdot) : \mathbb{R}^n \to \mathbb{R}^m$, $g(\cdot) : \mathbb{R}^n \to \mathbb{R}^r$, and $h(\cdot) : \mathbb{R}^n \to \mathbb{R}^p$. The functions g_i and h_j are the inequality constraint and equality functions, respectively. The vectors $x \in \mathbb{R}^n$ are called vectors of parameters of the multi-objective problem and form the parameter space. The vectors $f(x) \in \mathbb{R}^m$ are in a vector space called objective space.

Multi-objective optimization basically seeks to find a set of solutions, wherein each solution belonging to this set can not have one of its elements improved without loss of quality of the other objectives, or even without violation of the problem restrictions [237]. This set of solutions is called Pareto-optimal.

Formally, a solution $x^1 \in X$ dominates another solution $x^2 \in X$ if $f_i(x^1) \preceq f_i(x^2)$ and $f_i(x^1) \neq f_i(x^2)$ for all $i = 1, ..., m$. Likewise, it is said that there is a $j = 1, ..., m$ where $f_j(x^1) \in Y$ dominates $f_j(x^2) \in Y$, in these same conditions.

Therefore, a solution $x^* \in F_x$ is a Pareto-optimal solution if x^* is not dominated by any other feasible point. Thus, $X^* \subset X$ is a Pareto-optimal if all solutions that comprise it are Pareto-optimal solutions. The image set $Y^* \subset Y$ associated with the Pareto-optimal is called the Pareto-optimal front.

Provided that the functions of the problem to be optimized have been defined, i.e., fragility, moisture, and cost of production, two techniques based on evolutionary computation, NSGA-II (Nondominated Sorting Genetic Algorithm) [69] and SPEA2 (Strength Pareto Evolutionary Algorithm) [268], were chosen to perform the multi-objective optimization of the process.

Despite having three objective functions, two of them could be grouped due to presenting a strong correlation, i.e., the functions fragility and moisture, related to quality of the product, could be grouped using the weighted sum method. This method reduces the size of a multi-objective problem [167]. However, one limitation of the weighted sum method is the difficulty in setting the weights of each objective [136]. Equation 1.7 shows the formulation of the general problem.

$$\begin{cases} min : \lambda f_1 + (1 - \lambda) f_2 \\ min : F_{cost} \\ s.a. : f_1 < 0.50\% \\ f_2 < 3\% \\ where : x \epsilon X. \end{cases} \tag{1.7}$$

In Equation 1.7, the parameter λ indicates the relevance of each one of the objective functions in the optimization process, and the objective functions f_1 and f_2 represent moisture and electrode fragility, respectively.

During the experiments two values were adopted for λ, 0.2 and 0.50, generating two optimization problems that can be expressed by Equations 1.8 and 1.9.

$$\begin{cases} min : 0.25 f_1 + 0.75 f_2 \\ min : F_{cost} \\ s.a. : f_1 < 0.50\% \\ f_2 < 3\% \\ where : x \epsilon X. \end{cases} \tag{1.8}$$

The fragility presents a relevance three times higher than moisture in the problem represented by Equation 1.8. This argument was made considering that the electrode fragility problems occur more often than those related to moisture.

$$\begin{cases} min : 0.50f_1 + 0.50f_2 \\ min : F_{cost} \\ s.a. : f_1 < 0.50\% \\ f_2 < 3\% \\ where : x \epsilon X. \end{cases} \tag{1.9}$$

On the other hand, in the problem represented by Equation 1.9, both fragility and humidity presented the same value of relevance. This weighting was carried out in order to achieve a balance of quality parameters.

Computational experiments were performed in 20 runs for each one of the techniques SPEA2 and NSGA-II and for each value of λ. The parameters adopted for the implementation and execution of the evolutionary algorithms were chosen empirically in order to find the best results in a shorter runtime.

TABLE 1.9: Parameters of the evolutionary algorithms

Test	SPEA2	NSGA-II
Population	100	100
Size of the File	100	100
Number of Generations	500	500
Crossing Probability	0.8	0.8
Mutation Probability	0.1	0.1
Crossing Type	*One-point*	*One-point*
Mechanism of Niche	*Crowding*	*Crowding*

The hypervolume metric was used to evaluate the performance of each algorithm [268]. This measure allows to examine convergence, spreading, and coverage of a Pareto-optimal solution [268].

1.4.2 Experimental Tests

Figures 1.3 and 1.4 show the Pareto fronts obtained by the combination of all tests carried out to $\lambda = 0.25$ and $\lambda = 0.50$, respectively. The hypervolume metric was used to evaluate the performance of each algorithm. Table 1.10 shows the average results of the metric hypervolume for $\lambda = 0.25$ and 0.50. However, it is not possible to state which algorithm presents the best performance without a statistical verification of the relevance and representativeness of the results.

In order to evaluate which algorithm showed the best result we proposed to perform an analysis of variance (ANOVA) for all results, considering each value of λ. After verifying the initial hypothesis of normality, homoscedasticity,

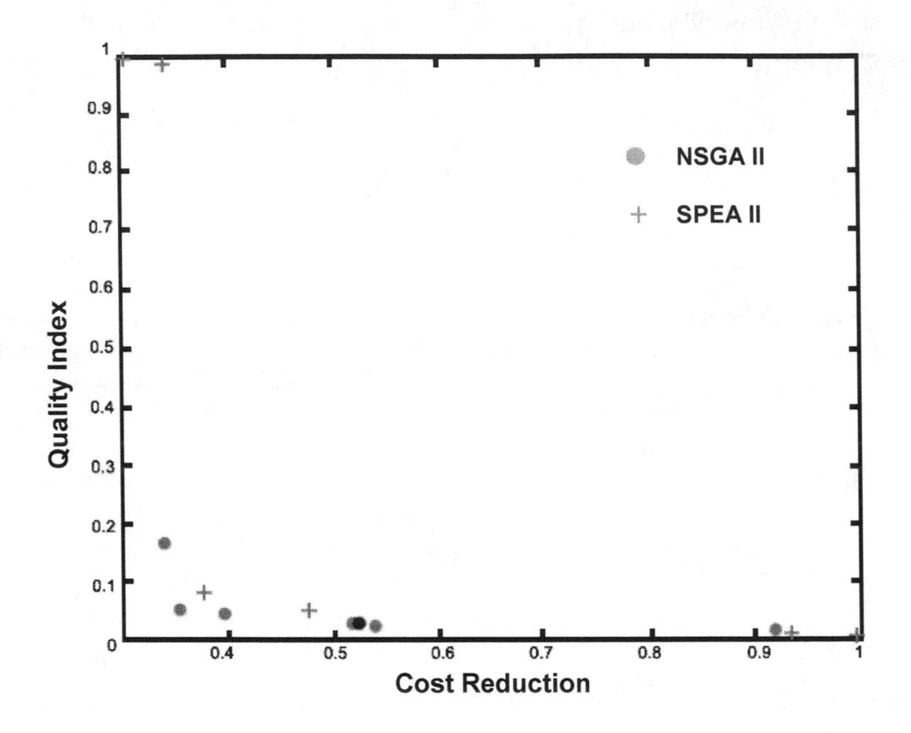

FIGURE 1.3: Pareto front ($\lambda = 0.25$).

and randomness of the samples, requisites required for applying ANOVA, tests were performed and results can been seen in Tables 1.11 and 1.12.

The analysis of variance (ANOVA) suggests that the algorithms showed different performances from a statistical point of view, i.e., NSGA-II presented the best performance for $\lambda = 0.50$ because *F-Value* is greater than *P-Value* [179]. On the other hand, for $\lambda = 0.25$, there is not statistical evidence that differentiate both algorithms, since the *F-Value* is smaller than *P-Value.*

Thus, in order to analyze the feasibility of implementation in a real production process two solutions obtained by the NSGA-II algorithm, one for each value of λ, were chosen. Table 1.13 displays the selected solutions.

The selection criterion of these solutions was based on the operating viability, i.e., it was based on the similarity of the solutions obtained with the real production process, since the operators are familiar with the current process. These selected solutions indicate a reduction of 22.4% and 31% in manufacturing costs for $\lambda = 0.25$ and $\lambda = 0.50$, respectively. These reduction values were calculated based on the current manufacturing process that presents a cost of US$0.82 per kilo of produced electrodes.

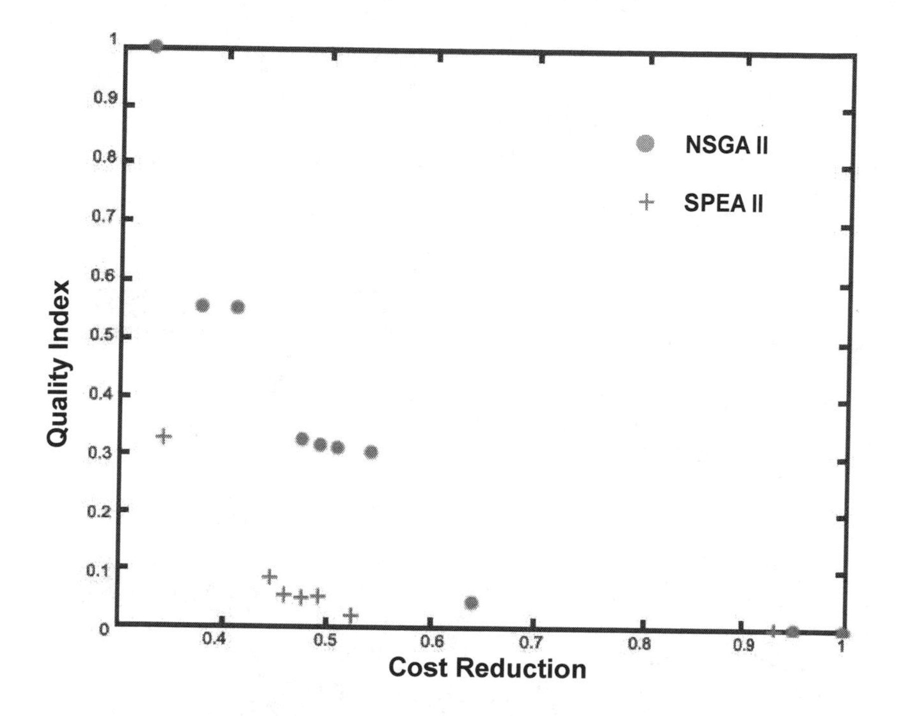

FIGURE 1.4: Pareto front ($\lambda = 0.50$).

TABLE 1.10: Hypervolume metric

Test	SPEA2	NSGA-II
P01 ($\lambda = 0.25$)	0.63	0.55
P01 ($\lambda = 0.50$)	0.54	0.59

TABLE 1.11: ANOVA test for $\lambda = 0.25$

	DF	SS	MS	F-Value	P-Value
Fator	1	0.0016	0.0016	0.06	0.807
Error	38	0.98	0.026		
Total	39	0.98			

1.5 Final Remarks

One purpose of this study was to propose the creation of a mathematical model that represents the manufacturing process of stainless steel coated elec-

TABLE 1.12: ANOVA test for $\lambda = 0.50$

	DF	SS	MS	F-Value	P-Value
Fator	1	0.023	0.023	1.73	0.196
Error	38	0.50	0.013		
Total	39	0.52			

TABLE 1.13: New production parameters

	T_{Ar}	t_1(h)	T_1(°C)	t_2(h)	T_2(°C)	f_1	f_2
Case 1	6h	1h	150 °C	3h	420 °C	0.14	US$0.64
Case 2	2h	2h	140 °C	3h	360 °C	0.17	US$0.57

trodes, seeking to understand the relationships present in each stage of manufacturing. Initially, measurements and statistical analysis were performed with the main variables that comprise the production process.

Once the process variables were defined, a model based on artificial neural network was implemented in two distinct platforms in order to find what model would be the most suitable for a real manufacturing process. After training and testing the models, it was verified that the model implemented in the JustNN platform showed better results than the model implemented in MATLAB. Now, taking into consideration the JustNN platform, three patterns of the process were chosen and tested in a real production system. The error obtained between the real values and simulated cases was less than 20%, and although the high magnitude, this result may be considered appropriate for a real manufacturing process. The computational model was capable of simulating the process at low cost with regard to time and material resources and was used as a basis in another objective of this work, i.e., optimization of the process.

The second goal, optimization of stainless electrode production process, took into consideration the parameters of quality, moisture, and fragility, as well as the production cost. The multi-objective optimization methods NSGA-II and SPEA2 were selected and objectives relating to product quality, moisture, and fragility were aggregated into a single function, using the weighted sum method. For this work, two values for the weighting factor, $\lambda = 0.25$ and $\lambda = 0.50$, were used in the aggregation of the objectives. Computational experiments were performed in 20 runs for each one of the techniques SPEA2 and NSGA-II and for each value of λ. Finally, the performance of all tests was evaluated using the hypervolume metric.

The results showed the sensitivity of the multi-objective algorithms regarding the parameter λ used in the aggregation of the variables, moisture, and fragility of the product. It was observed by ANOVA that for $\lambda = 0.25$ no statistica evidence was found that differentiate the NSGA-II and SPEA2 algorithms. However, for $\lambda = 0.50$ the NSGA-II algorithm presented better performance than SPEA2. As a consequence, the algorithm chosen to validate the method was NSGA-II.

In order to investigate the validity of the method and the feasibility of implementation in a real production process, two solutions obtained by the NSGA-II algorithm were chosen. Finally, the results showed that it is possible to find new parameters for the production process that, in addition to complying with the quality requirements, achieved a reduction in the manufacturing costs of 22.4% and 31% for values of $\lambda = 0.25$ and $\lambda = 0.50$, respectively.

Finally, considering the statistical analyzes as well as the relative difference in the reduction of production costs obtained for different values of λ, a suggestion for future work would be to conduct the analysis of the production process taking into account three objectives, i.e., without performing the aggregation of the moisture and fragility as a single goal.

Acknowledgements

The authors are thankful for, the support of CAPES-Brazil, CNPq-Brazil, FAPEMIG, and CEFET-MG.

Chapter 2

Learning Fuzzy Rules from Imbalanced Datasets using Multi-objective Evolutionary Algorithms

Edward Hinojosa Cárdenas

National University of San Agustín, Arequipa, Peru

Heloisa A. Camargo

Federal University of São Carlos, São Carlos, São Paulo, Brazil

Yván Jesús Túpac Valdivia

San Pablo Catholic University, Arequipa, Peru

2.1 Introduction

Computational intelligence is a research area oriented to model different aspects of intelligence embracing methodologies such as neural networks, fuzzy systems, and evolutionary algorithms. The main characteristic of computational intelligence is to explore the combination of two or more methodologies that cooperate with each other to improve results and enhance effectiveness and robustness of the resulting system. One of the most successful hybrid approaches in computational intelligence is the use of evolutionary algorithms, specially the Genetic Algorithms (GAs), to generate and tune fuzzy systems. From this type of combination emerged the Genetic Fuzzy Systems (GFSs) [111] [112].

Fuzzy Rule Based-Systems (FRBSs) are one of the most important applications of fuzzy set theory [249]. Generally speaking, they present two components: the Knowledge Base (KB) and the inference mechanism. The KB contains the knowledge about the domain of interest used by the system to make inferences and solve the problem. The KB is usually formed by two components: a set of fuzzy rules, called fuzzy Rule Base (RB) and the fuzzy partitions, containing the fuzzy sets for each variable appearing in the fuzzy rules, called the Data Base (DB). The inference mechanism implements the reasoning method used by the system and generates an output when an input is specified. FRBSs have been developed to solve different types of problems such as modelling, control, and classification. FRBSs that focus on classification problems are called Fuzzy Rule-Based Classification Systems (FRBCSs).

In the field of GFSs, the possible ways of using GAs to generate Fuzzy System (FS) from data are broadly classified as learning and tuning processes. In a learning process, the DB, the RB, or both are automatically generated from scratch. In the tuning process, one of the components of the systems that have been generated before is adjusted to improve the ability of the system to find accurate results. Since one of the main advantages of representing knowledge by means of fuzzy rules is the possibility of understanding the rule meaning, and why the system generated one specific output, the FS is expected to present both good accuracy and good interpretability, which are conflicting objectives. Towards this goal, researchers have explored, in recent years, the use of Multi-Objective Evolutionary Algorithms (MOEAs), since in these algorithms, the search for solutions that balance conflicting objectives is embedded in the algorithm itself. The term Multi-Objective Evolutionary Fuzzy System (MOEFS) was coined to refer to the research field dedicated to learning FS from data using MOEA [85].

According to [85], learning classification fuzzy systems from imbalanced datasets is one of the research trends in MOEFS. Imbalanced datasets are characterized by having very different numbers of examples in each class. In the typical binary class imbalanced problem, one class (negative or majority class) vastly outnumbers the other (positive or minority class) and the positive class is usually the one that has the highest interest [113]. In this context, the majority class influences the learning process of the classifiers that obtain generally a good accuracy on the negative class but very poor accuracy on the minority class because most of the traditional algorithms pursue to minimize the general error rate [95]. Furthermore, other problems have to be considered by the algorithms that generate classifiers from imbalanced datasets, such as small disjunct and overlapping classes, among others. The solutions to the imbalanced datasets problem have been proposed by means of techniques that can be grouped in different levels, according to the phase of the whole learning process to which the technique is applied.

This work presents a proposal to generate fuzzy rule-based classification systems from imbalanced datasets, using multi-objective genetic algorithms based on the Iterative Rule Learning (IRL) approach. The proposed method, called Iterative Rule Learning from Imbalanced Datasets using MOEA (IRL-ID-MOEA), includes two phases: preprocessing and rules generation. Each one of these phases provides techniques that focus on different levels of the imbalanced datasets problem: the data level and the feature selection level. The preprocessing phase consists of running an algorithm that generates a balanced dataset from the original imbalanced one, to be used in the next phase. This process clearly consists of a data level technique to the imbalanced dataset problem. The method described here has been described in a short version in [32]. In this work, the presentation has been expanded with the inclusion of experiments and analysis with ten preprocessing methods to balance the datasets.

The rule generation phase begins with the definition of fuzzy partitions for each variable and is based on the IRL approach, where each chromosome represents a fuzzy rule. The MOEA is run several times and, at each time, the best rule is selected and inserted in the RB. The process is repeated until no improvements are obtained with the insertion of new rules in the RB. The rule generation phase includes a technique in the feature selection level, for it allows the representation of don't care conditions in the chromosomes. The MOEA used in this work is the well-known Non-dominated Sorting Genetic Algorithm (NSGA-II).

This chapter is organized as follows. In Section 2.2, we present the imbalanced dataset problem and describe the oversampling, undersampling, and hybrid methods used in this work. In Section 2.3, basic concepts of Fuzzy Rule-Based Systems and Fuzzy Rule-Based Classification Systems are presented. The Genetic Fuzzy Systems are described in Section 2.4. The method proposed for preprocessing the imbalanced dataset and learning fuzzy classification rules using the IRL approach and MOEA is described in Section 2.5. The experimental analysis comprising ten different preprocessing algorithms is presented in Section 2.6. Finally, some concluding remarks and future works are presented in Section 2.7.

2.2 Imbalanced Dataset Problem

Several real-world problems are characterized by imbalanced learning data. Examples include (but are not limited to) face recognition [155], remote sensing [243], level ozone prediction [235], and medical diagnosis [96].

In this work, we consider the problem of the binary imbalanced dataset, where there are two classes, the positive or minority class and the negative or majority class. The main characteristic of this type of database is the significant difference between the number of examples of each class. We can consider that a difference is significant when it is at least 9:1, i.e., the number of examples of the majority class is at least nine times the number of examples of the minority class. We can represent that difference by the Imbalanced Ratio (IR). The IR is defined as the ratio of number of instances in majority class to the number of instances in minority class [164].

The significant difference is a problem in the automatic generation of classifiers, because the model generated by the traditional learning methods does not present a good performance when classifying examples of positive class (class with less examples) [162]. Usually the minority class is the most important one, for example, in medical diagnosis of a cancer disease [95].

In addition to the significant difference, methods for automatically generating classifiers on imbalanced databases face other problems, among which we can highlight small disjuncts [129] and overlapping problems [203]. Small disjuncts are small subsets of one class that are separated from the largest subset of the same class. Overlapping occurs when the imbalanced dataset contains ambiguous regions in the data space where the prior probability of two or more classes are approximately equal. These two problems are illustrated in Figure. 2.1 [89].

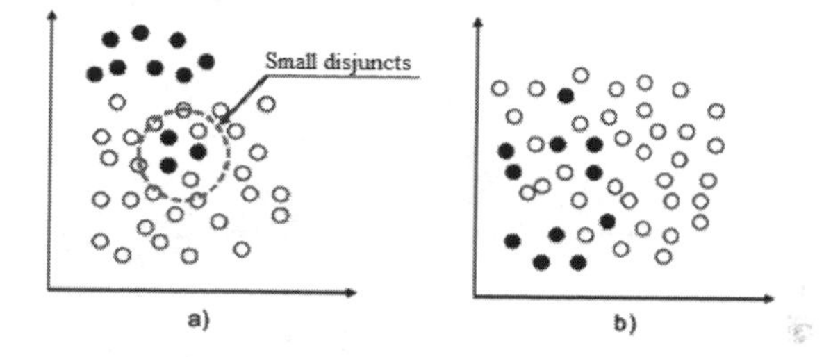

FIGURE 2.1: (a) Small disjuncts (b) Overlapping [89].

There are different techniques for dealing with imbalanced classification problems and they can be grouped in five levels: data level, algorithm level, cost sensitive level, feature selection level, and ensemble level [208].

Different works (for example, [208], [157], and [95]) considered that the pre processing methods in the data level provide better results than the other methods and techniques focusing on other levels, but they suggest to use two or three methods or techniques dealing with different levels during the training process. In this work, we propose and use techniques that focus on two levels, data level and feature selection level. In the data level we use well-known methods of oversampling, undersampling, and hybrid strategies. They are detailed in the next subsections. In the feature selection level we use an approach that generates fuzzy classification rules from data with don't care conditions (Section 2.5).

2.2.1 Oversampling Methods

In the learning process of a classifier on imbalanced dataset, the oversampling method is used to add new minority class examples in order to balance the class distributions.

One of the first oversampling techniques proposed consists in repeating positive examples as many times as necessary as shown in Figure 2.2 [157].

This technique doesn't add new information in the learning process and, for this reason, it has not been used. The most commonly used oversampling techniques add new positive examples called synthetic examples in order of generating variation in the information that helps the learning process of the classifier [36]. In the next subsections, some of the well-known oversampling techniques that are used in the proposed method are listed and briefly explained.

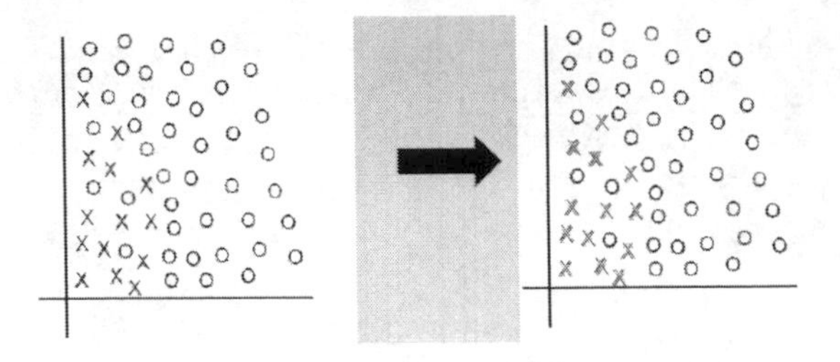

FIGURE 2.2: Repetition of positive examples [157].

2.2.1.1 Synthetic Minority Over-sampling TEchnique (SMOTE)

SMOTE was proposed in [42]. It focuses on oversampling the minority class, creating synthetic examples by performing certain operations on the values of real selected examples.

The minority class is oversampled adding synthetic examples along the line segments that connect a selected real example to one or all k nearest neighbors of the minority class. Depending on the amount of synthetic examples required, a subset of the k real nearest neighbors are chosen at random. For example, if the amount of oversampling required is 200%, only two neighbors of the k real nearest neighbors are selected and examples are generated in their direction.

Briefly, synthetic examples are generated as follows: select a real example, calculate the distance between this example and other examples, and select k real nearest neighbors. For each real nearest neighbor find the line segment that unites the selected real example to its neighbor, and, finally, selects a random point on the segment found as the new synthetic example. The selection of a random point along the line segment between two specific real examples assures that the decision region of the minority class becomes more general. Figure 2.3 shows the generation process of a synthetic example using SMOTE.

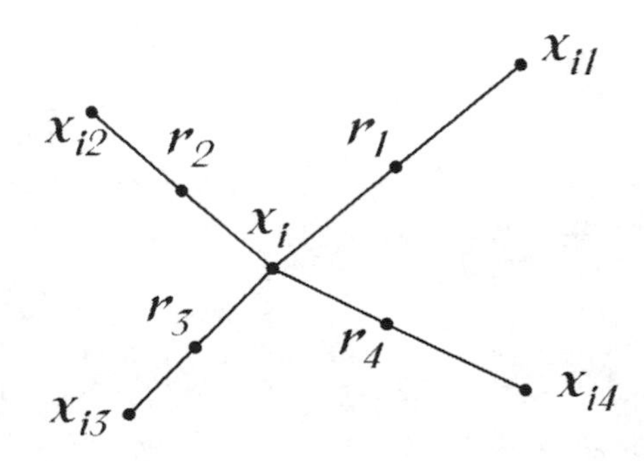

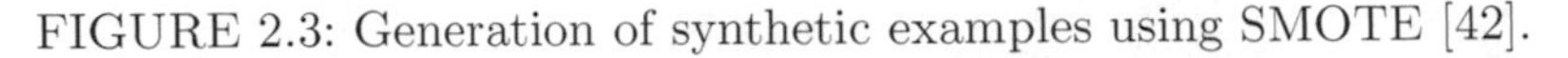

FIGURE 2.3: Generation of synthetic examples using SMOTE [42].

2.2.1.2 Borderline-Synthetic Minority Over-sampling TEchnique (Borderline-SMOTE)

Borderline-SMOTE [106] is based on SMOTE. The main difference with relation to SMOTE is that Borderline-SMOTE generates synthetic examples only from real examples of the minority class on the border of regions containing minority examples.

Figure 2.4 shows an example Borderline-SMOTE. Part (a) shows the original distribution of the imbalanced database, where circular points represent the majority class and cross points represent the minority class. Part (b) shows examples of the minority class that are selected to generate the synthetic examples. The selected examples represented by square blue points have a similar number of positive and negative examples among their k nearest neighbors. Part (c) shows the synthetic examples generated from select real examples using SMOTE, which are represented by square points without filling.

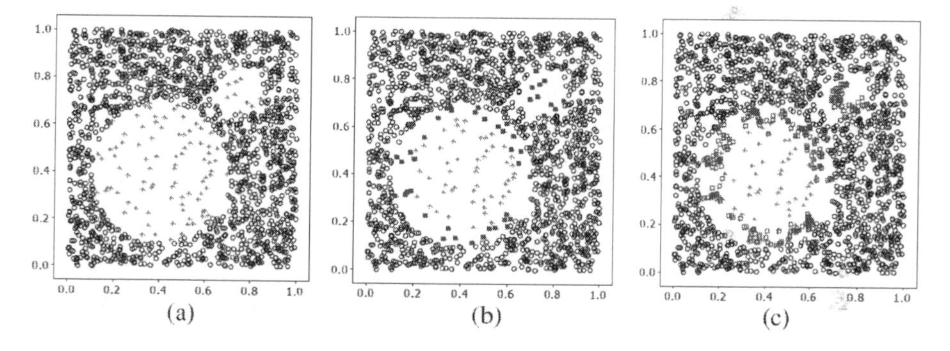

FIGURE 2.4: (a) Original imbalanced database. (b) Real minority examples in the border. (c) Synthetic examples generated [106].

2.2.1.3 ADASYN (ADAptive SYNthetic Sampling)

ADASYN was proposed in [110]. This technique generates synthetic examples of the minority class adaptively according to the distribution of the examples in the classes. More synthetic examples are generated for the minority class examples that have more Negative class examples among their k nearest neighbors, which are the examples that are more difficult to learn. Figure 2.5 shows an oversampling procedure using ADASYN.

2.2.1.4 Safe-Level-SMOTE (Safe Level Synthetic Minority Oversampling TEchnique)

Safe-Level-SMOTE was proposed in [31] and it's based on SMOTE. This technique defines if a positive example is safe before generating synthetic examples from that positive example. Each synthetic example is generated closer to safe positive examples.

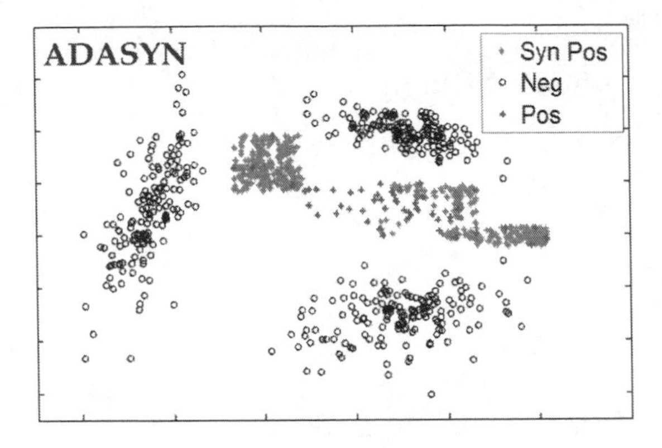

FIGURE 2.5: Oversampling using ADASYN [36].

The safe level *sl* of a positive real example is defined by the amount of positive examples among the K nearest neighbors of this example. If *sl* of a positive example is close to 0, it can be considered a noise; if *sl* is close to K the example is considered safe. The safe level ratio is used to select the position of a synthetic example and is defined by dividing the *sl* of a positive example by the *sl* of its nearest neighbor.

2.2.2 Undersampling Method

The undersampling methods are also used to balance the amount of minority and majority class examples by reducing or removing examples of the majority class.

One of the first undersampling techniques found consists in removing negative examples randomly until the amount of examples are balanced as shown in Figure 2.6 [157].

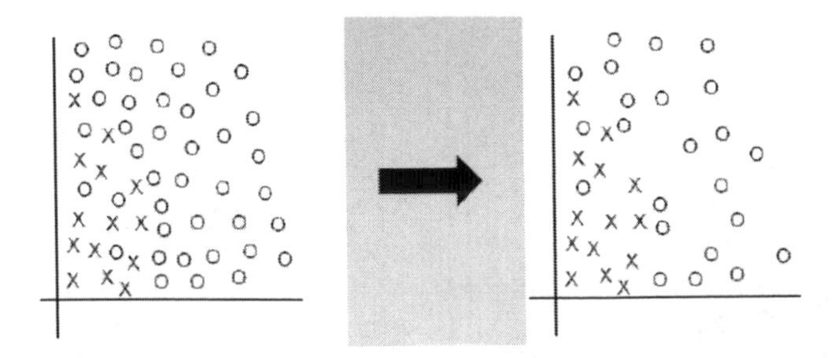

FIGURE 2.6: Removing negative examples randomly [157].

When this technique is used, important information can be lost and the classifiers' performance can be reduced. The most commonly used undersampling techniques reduce or remove negative examples considered redundant or noisy [95]. In the next sections some well-known undersampling techniques that are used in the proposed method are listed and briefly explained.

2.2.2.1 TL (Tomek Link) Technique

TL technique was proposed by Ivan Tomek in 1976 [233]. This technique selects and removes negative examples in the border or negative examples considered noise. TL technique follows the following steps:

1. Select two examples e_i and e_j from different classes.
2. The distance between those examples is defined as $dist(e_i, e_j)$.
3. The pair (e_i, e_j) is called TL if there isn't another example e_k closer to e_i or e_j, i.e., $dist(e_i, e_j) < dist(e_i, e_k)$ and $dist(e_i, e_j) < dist(e_j, e_k)$.
4. The negative example (e_i or e_j) is removed from the dataset.

Figure 2.7 shows the process of TL technique. The negative examples are represented by circular dots.

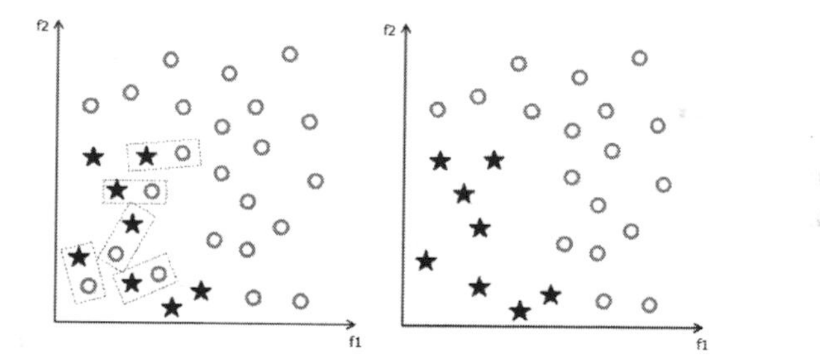

FIGURE 2.7: Example of Tomek link technique [59].

2.2.2.2 OSS (One Sided Selection) Technique

OSS was proposed in [143]. It's an undersampling technique that keeps only the most representative examples of the majority class.

This technique of selecting and removing examples consists of the application of TL followed by CNN (Condensed Nearest Neighbor). TL is used as a method of selection of instances that removes noise and border examples from the majority class. CNN is used to remove the majority class examples that are distant from the decision border.

2.2.2.3 NCL (Neighborhood CLeaning Rule) Technique

NCL was proposed in [146]. This technique uses the ENN (Edited Nearest Neighbor rule) [245] to remove examples. For each example, the three nearest neighbors are found. If the example belongs to the majority class and its neighbors to the minority class, the original example is removed. If the example belongs to the minority class and its neighbors belong to the maiority class, the neighbors are removed. Figure 2.8 shows the NCL algorithm.

Require: L: a list of examples
Require: T: training data set
1. **for all** e_i in T **do**
2. VNN = 3-NN of e_i, in T
3. **if** $e_i \in$ majorityclass **then**
4. **if all** e_j in VNN $\in$ minorityclass **then**
5. insert e_i into L
6. **end if**
7. **end if**
8. **if** $e_i \in$ minorityclass **then**
9. **if all** e_j in VNN $\in$ majorityclass **then**
10. insert **all** e_j in VNN into L
11. **end if**
12. **end if**
13. **end for**
14. $T = T - L$
15. **return** T

FIGURE 2.8: NCL algorithm [146].

2.2.2.4 SBC (underSampling Based on Clustering) Technique

SBC was proposed in [146]. This technique groups all examples in some groups and then selects a number of negative examples to be eliminated. The selection is based on the ratio between the amount of negative examples and the amount of positive examples in each group, i.e., negative examples in groups with a higher ratio (between the amount of negative examples and the amount of positive examples) will have more probability of being selected.

2.2.3 Hybrid Methods

2.2.3.1 SMOTE + TL

The SMOTE + TL hybrid technique is a combination of the oversampling technique SMOTE and the undersampling technique Tomek Link. An example of the SMOTE + TL hybrid method is shown in Figure 2.9 [16]. First, positive examples are oversampled using SMOTE. In the sequel, examples of the negative class are removed using TL.

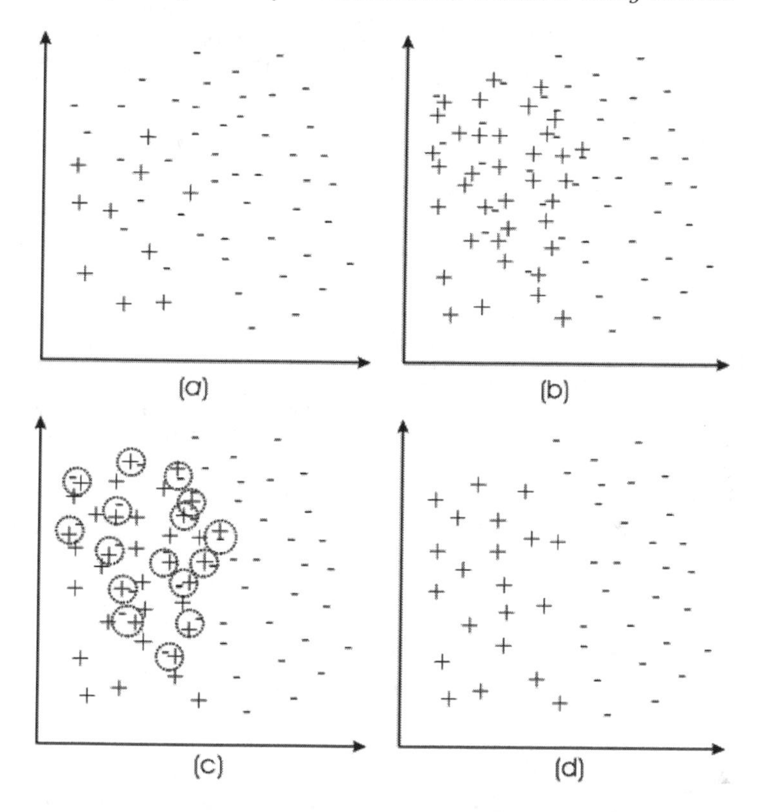

FIGURE 2.9: SMOTE+TL method: (a) Original imbalanced dataset. (b) Oversampling with SMOTE. (c) Select TL pairs. (d) Remove negative examples [16].

2.2.3.2 Synthetic Minority Over-sampling TEchnique + Edited Nearest Neighbor

SMOTE + ENN was proposed in [16] too and it's similar to the SMOTE+TL technique. The difference is that this technique uses ENN instead of TL for removing positive and negative class examples. In ENN, if an example is incorrectly classified based on its 3 Nearest Neighbors (3-NN) then this example is removed.

2.2.4 Evaluation Measure for Classification in Imbalanced Datasets

Usually, the classification rate is used to determinate the accuracy of the classifier, but in the imbalanced classification problem that rate may be influenced by a number of negative examples or majority class.

For this, we use a well-known measure for binary classification called area under the ROC curve (AUC) [118] to determine the accuracy of our classifier and to compare it with other methods of learning fuzzy rules for the imbalanced classification problem. We explain this measure in the next subsection.

2.2.4.1 Area under the ROC Curve

The AUC is a standard method that defines the performance of a classifier and it's based on the following values:

1. False Positive (FP): The number of negative examples classified as positive examples.
2. False Negative (FN): The number of positive examples classified as negative examples.
3. True Positive (TP): The number of positive examples classified as positive examples.
4. True Negative (TN): The number of negative examples classified as negative examples.

The AUC value is defined by the next equations:

$$AUC = \frac{1 + TP_{rate} - FP_{rate}}{2}. \tag{2.1}$$

$$TP_{rate} = \frac{TP}{TP + FN}. \tag{2.2}$$

$$FP_{rate} = \frac{FP}{TN + FP}. \tag{2.3}$$

For example, we calculate the AUC of a classifier for a simple problem in a banking organization. It has 1000 clients that request a bank credit; 300 are reliable and 700 are unreliable. Assuming that the results of the classifier are $TN = 100$, $FN = 100$, $FP = 200$, and $TP = 600$, we define the values $TP_{rate} = 600/(600 + 100) = 0.86$ and $FP_{rate} = 200/(100 + 200) = 0.67$. Finally, we calculate $AUC = (1 + TP_{rate} - FP_{rate})/2 = (1 + 0.86 - 0.67)/2 = 0.60$. Figure 2.10 shows the above mentioned example.

2.3 Fuzzy Rule-Based Systems

Fuzzy systems are based on fuzzy set theory and fuzzy logic. Fuzzy systems was proposed by Lofti A. Zadeh in 1965 [249] to resolve problems with vague, imprecise, and incomplete information. An example for each kind of information about a meeting is shown following:

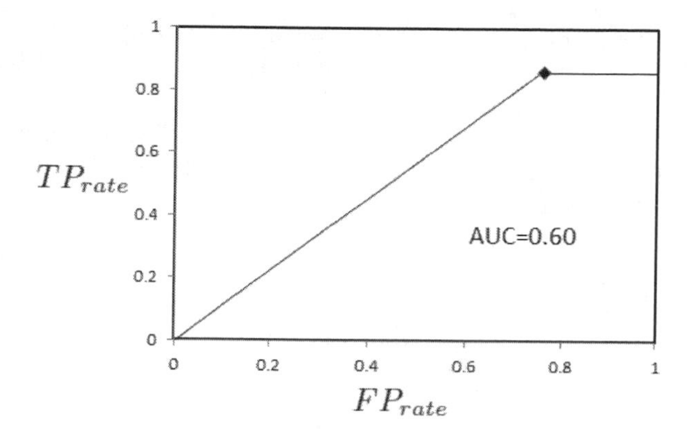

FIGURE 2.10: Example of AUC.

- Vague information: The meeting is more or less at 8 a.m.
- Imprecise information: The meeting is between 8 a.m. and 9 a.m.
- Incomplete information: I don't know when is the meeting, but usually it starts at 8 a.m.

The FRBSs are a prominent type of fuzzy systems that use fuzzy set theory and fuzzy logic to infer conclusions from vague, imprecise, and incomplete information.

Generally, a FRBCS is formed by two main parts [32]: a knowledge base and an inference mechanism.

1. The Knowledge Base (KB) is composed of a Fuzzy Rule Base (FRB) and a Fuzzy Data Base (FDB). The FRB contains a set of fuzzy rules that form the core of the system. The FDB includes the definition of the attributes (linguistic variables), also called features, in terms of fuzzy sets (linguistic terms).
2. The Inference Mechanism (IM) uses the information of the FRB and FDB to infer new information. In the case of FRBS for classification, the problem of interest for this work, the purpose of the IM is to classify new examples.

In the next subsection, the usual form of fuzzy classification rules as well as the inference methods for classification are explained in more detail.

2.3.1 Fuzzy Rule-Based Classification Systems

In many real problems the classification is an important task, for example, data mining, sentiment analysis, computer vision applications, among others. Briefly, the classification process consists of, given a set of examples

$E = \{e_1, e_2, ..., e_p\}$, to assign a class C_j of a set of classes $C = \{C_1, C_2, ..., C_m\}$ for each example $e_q \in E$, where each e_q is defined by n features $e_q = \{a_{q1}, a_{q2}, ..., a_{qn}\}$.

In these works we use FRBS to perform the classification task. These systems are called Fuzzy Rule-Based Classification Systems. The main difference between a FRBS and FRBCS is that in the FRBCS the fuzzy rules have a class as a consequent. A typical fuzzy classification rule can be expressed by:

$$R_i\text{: } \textbf{IF } V_1 \textbf{ IS } T_{1l_1} \textbf{ AND } V_2 \textbf{ IS } T_{2l_2} \textbf{ AND } ... \textbf{ AND } V_n \textbf{ IS } T_{nl_n}$$
$$\textbf{THEN } \text{Class} = C_j$$

where:

R_i	: Fuzzy rule with identifier i.
$V_1, V_2, ..., V_n$	: Linguistic variables or features of the set of examples considered in the problem.
$T_{1l_1}, T_{2l_2}, ..., T_{nl_n}$	: Linguistic terms used to represent the feature values.
C_j	: It's the class of the R_i.

The most used and well-known methods in the literature are classic and general fuzzy reasoning methods. We explain briefly the two methods in the next subsections. This work uses the classic fuzzy reasoning method.

2.3.2 Classic Fuzzy Reasoning Method

This method classifies an example using the rule that has the highest compatibility degree with the example. To define the class $C_j \in C = \{C_1, C_2, ..., C_m\}$ of an example e_q represented by n features $e_q = \{a_{q1}, a_{q2}, ..., a_{qn}\}$ this method applies the following steps:

1. Calculate the compatibility degree between example e_q and all fuzzy rules in the RB. Generally, the evaluation of the compatibility degree uses a $t - norma$.

2. Find rule $R_i \in$ RB with the highest compatibility degree with the example e_q.

3. Assign the class C_j (consequent of class R_i) to example e_q.

2.3.3 General Fuzzy Reasoning Method

This method classifies an example using an aggregation of the compatibility degrees of the rules that have the same class in the consequent. To define the class $C_j \in C = \{C_1, C_2, ..., C_m\}$ of an example e_q represented by n features $e_q = \{a_{q1}, a_{q2}, ..., a_{qn}\}$ this method applies the following steps:

1. Calculate the compatibility degree between example e_q and all fuzzy rules in RB. Generally, the evaluation of the compatibility degree uses a $t - norma$.

2. Calculate a classification value $Class_c$, for each class. $Class_c$ is defined as the aggregation of the compatibility degrees, calculated using all rules with class C_c in the consequent.

3. Assign the class C_j (with the highest classification value found in the previous step) to example e_q.

2.4 Genetic Fuzzy Systems

Genetic Fuzzy Systems (GFSs) are a soft computing paradigm which aims at building fuzzy systems using Genetic Algorithms (GAs). The term GFS emerged a few decades ago [52] and since then has been widely researched and numerous proposals can be found in the literature

Generally, GFS use GA to learn or to optimize parts of a FRBS (KB, FRB, FDB, or IM). In [112], the author classifies the GFS proposals according to a complete and comprehensive taxonomy. This classification is shown in Figure 2.11.

A detailed explanation of the categories in the classification can be found in [112], [111], and [53]. In the next subsections, we briefly explain the Genetic Rule Learning (a priori DB) and Multi-Objective Evolutionary Fuzzy Systems (MOEFS) that are of interest in this work, with emphasis on the types of coding used.

2.4.1 Genetic Rule Learning

Several works proposed a learning of RB from numeric information using an AG using a pre-defined DB. Figure 2.12 shows this process.

Generally, in genetic rule learning, four approaches are used. The first one is the Pittsburgh approach [224], where each chromosome defines a whole FRB and at the end of the genetic process the best chromosome is the final RB. Figure 2.13 shows the Pittsburgh approach.

The second approach is the Michigan approach [115], where each chromosome represents a rule and the RB is formed by the set of chromosomes in the best population. Figure 2.14 shows the Michigan approach.

The third approach is the iterative approach, where each chromosome represents a rule as in the Michigan approach. The difference between the iterative and Michigan approaches is that in this approach the GA is run several times and a new rule is adapted and added into RB at each execution of the

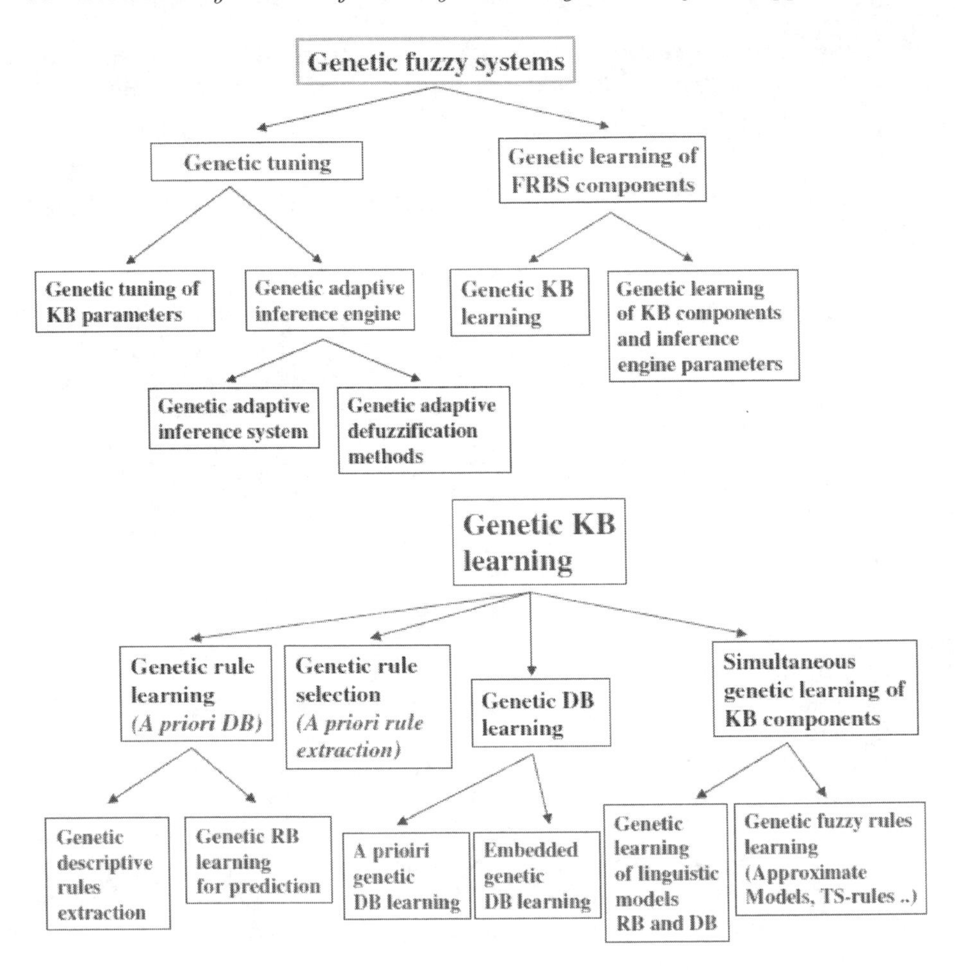

FIGURE 2.11: Classification of genetic fuzzy systems [112].

GA. The processes that use the iterative approach are called Iterative Rule Learning (IRL) [102]. The method proposed in this work is based on the IRL approach. Figure 2.15 shows the iterative approach.

Finally, the fourth approach is the Cooperative–Competitive approach [103], where each chromosome defines a whole RB like the Pittsburgh approach (Competitive), and each chromosome can improve based on the Michigan approach (Cooperative), i.e., it's a hybrid approach between the Pittsburgh approach and Michigan approach. Figure 2.16 shows the Cooperative–Competitive approach.

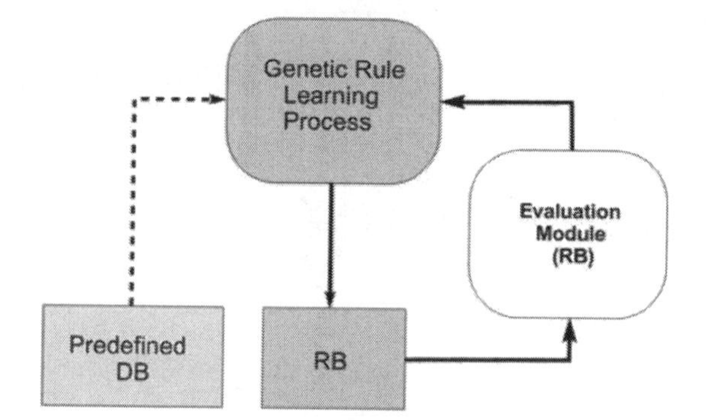

FIGURE 2.12: Genetic rule learning [112].

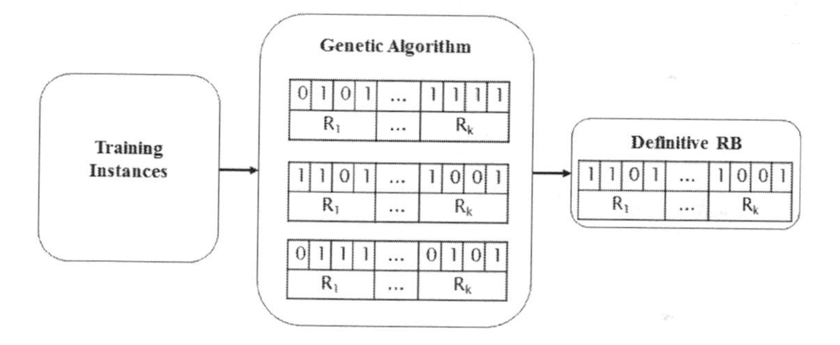

FIGURE 2.13: Pittsburgh approach.

2.4.2 Multi-Objective Evolutionary Fuzzy Systems

Multi-Objective Evolutionary Algorithms (MOEAs) are used in problems with multiple conflicting objectives, where the improvement of an objective leads to the deterioration of the others. Recently, MOEAs have been adopted as a more suitable optimization technique to generate fuzzy systems. These type of algorithms are appropriate for the task because they have an embedded process to balance conflicting objectives, which is the case with FRBS. Fuzzy rules are expected to be comprehensible and the FRB is expected to present a good performance. As these two objectives are contradictory, MOEA can be used to model the problem of learning FRBS in a natural way. The hybridization of fuzzy systems and MOEA is known as the Multi-objective Evolutionary Fuzzy Systems (MOEFSs) and in the taxonomy of MOEFS (Figure 2.17) this work is in the group to generate FRBS with different Acurracy–Interpretability trade-offs.

In this work we use a well-known MOEA, namely NSGA-II [62], to generate a set of non-dominated fuzzy classification rules. In the NSGA-II, the population Q_t (size N) is generated using the parent population P_t (size N). After this, the two populations are combined for generating the population R_t

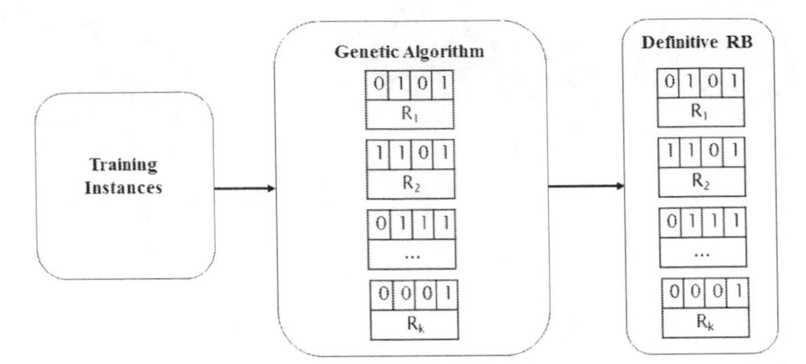

FIGURE 2.14: Michigan approach.

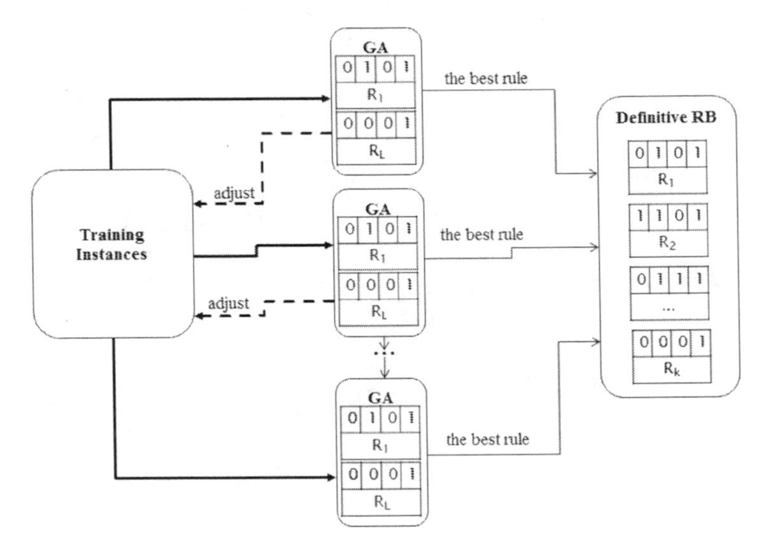

FIGURE 2.15: Iterative approach or IRL.

(size $2N$). The population R_t is sorted according to the dominance of the solutions in different Pareto fronts and the crowding distance. This distance is a measure of how close an individual is to its neighbors in the same front. Large average crowding distance will result in better diversity in the population.

A new population P_{t+1} (size N) is generated with the best Pareto fronts F_1, F_2, F_3, and so forth, until the P_{t+1} size equals to the value of N. The solutions in the Pareto fronts under this limit are removed. After P_{t+1} becomes the new P_t the process is repeated until a conditions is satisfied. The Figure 2.18 shows the process of evolution of the solutions in the NSGA-II.

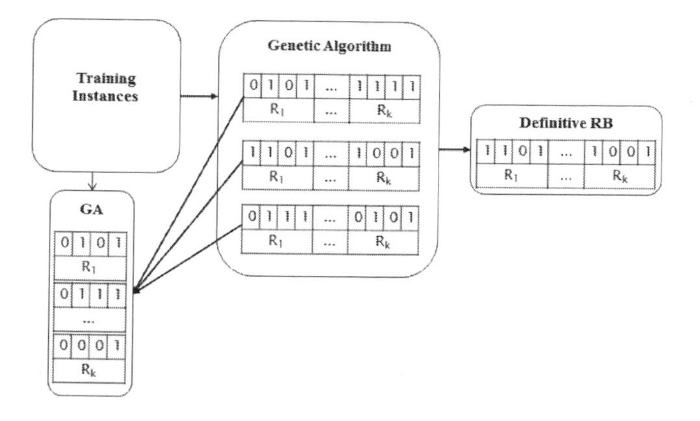

FIGURE 2.16: Cooperative–competitive approach.

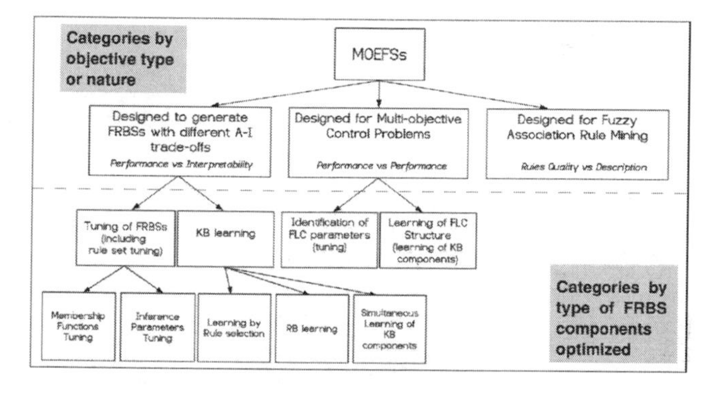

FIGURE 2.17: Two-level taxonomy based on the type of the objectives optimized (1st level) and on the type of GFS used (second level) [85].

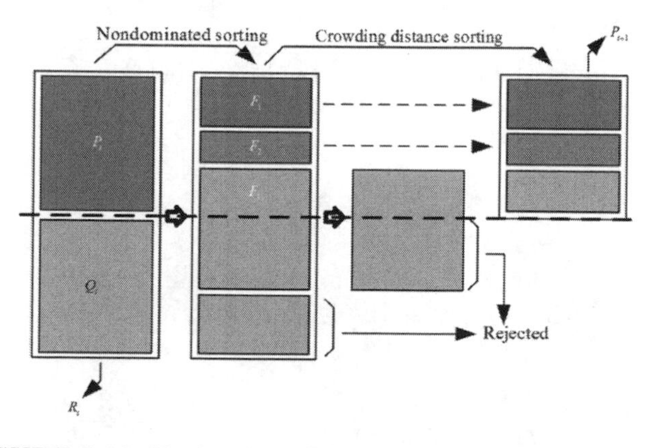

FIGURE 2.18: Evolutions of the solutions in the NSGA-II.

2.5 Proposed Method: IRL-ID-MOEA

In the context of FRBCS, several proposals that focus on the problem of imbalanced datasets can be found. Generally speaking, most of these proposals can be classified as cost-sensitive [158] [195] or preprocessing based [194] [238] [159] [161] approaches. The work in [90] proposes the genetic tuning based on 2-tupla representation and the one in [94] proposes a new fitness function, focused on imbalanced datasets. The work in [4] contains an experimental study of methods designed to deal with the imbalanced dataset problem. A features weighting method is proposed in [3] and a method to extract fuzzy rules from very large imbalanced datasets is proposed in [160].

The proposed method in this work to generate fuzzy classification rules, called Iterative fuzzy classification Rules Learning from Imbalanced Dataset by means of a Multi-Objective Evolutionary Algorithm (IRL-ID-MOEA), is presented in this section. The objective of this method is to generate fuzzy rules from imbalanced datasets and it is based on two phases as illustrated in Figure 2.19.

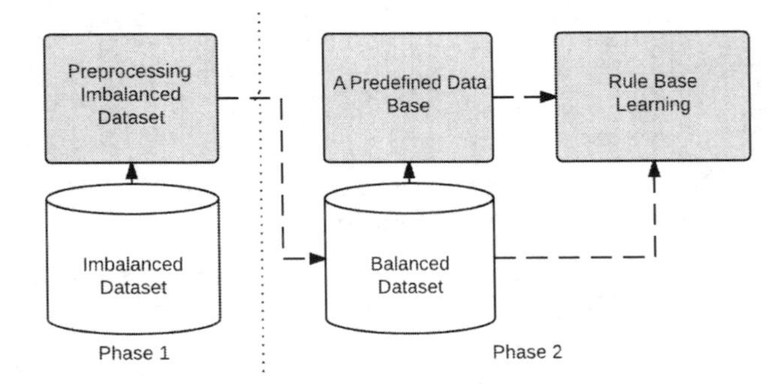

FIGURE 2.19: The IRL-ID-MOEA proposed method. Phase 1: Preprocessing the Imbalanced Dataset. Phase 2: Fuzzy Rules Learning.

In phase 1, an imbalanced dataset is preprocessed to be transformed in a balanced dataset. In this work the preprocessing techniques described in Section 2.2 have been used in the experiments.

In phase 2, the fuzzy classification rules are generated by means of an extension of the IRL based on a MOEA proposed in [37]. This phase consists of a number of steps detailed in Figure 2.20 and explained in the next sections.

2.5.1 A Predefined Dataset

The attributes of the dataset are represented by linguistic variables. In this step, the linguistic variables' domains are granulated by triangular and uniformly distributed fuzzy sets. Each linguistic variable has a number of

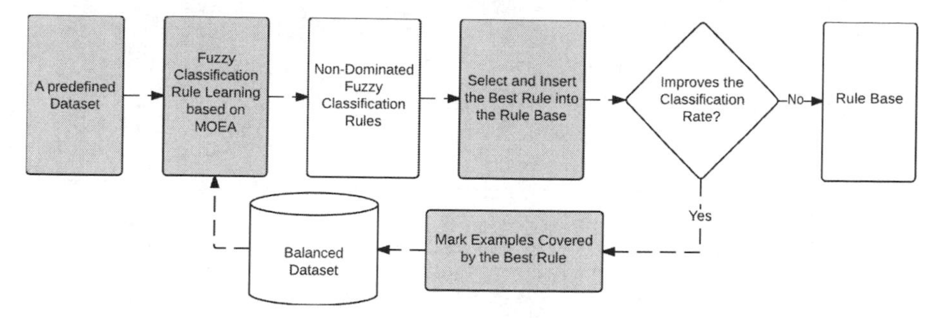

FIGURE 2.20: Iterative fuzzy classification rules learning based on a MOEA.

linguistic terms, defined as $\#LT$. The same number of fuzzy sets is used for every linguistic variable. An example of five fuzzy triangular sets is shown in Figure 2.21, where min is minimum value of linguistic variable, max is the maximum value of linguistic variable, and a_i, b_i, c_i are the parameters of each fuzzy triangular set.

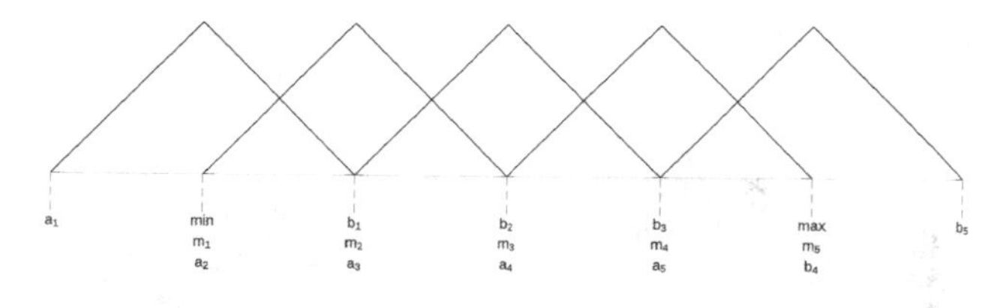

FIGURE 2.21: Uniformly distributed fuzzy sets.

2.5.2 Fuzzy Classification Rule Learning Based on MOEA

This work makes use of an IRL approach based on a MOEA to learn a set of fuzzy classification rules, i.e., a fuzzy classification rule is learned in each run of the MOEA. In the IRL approach, each chromosome encodes a fuzzy classification rule, similar to what is done in other works such as [1] [101], among others; each chromosome is represented by a vector of integer numbers where each element of the vector represents a gene. The length of the vector or number of genes depends on the number of linguistic variables considered in the problem, plus a gene that represents the class of the fuzzy rule. The first gene represents the index of the linguistic term or fuzzy set in the first linguistic variable, the second gene represents the index of the linguistic term or fuzzy set in the second linguistic variable, and so on; the last gene represents the class of the fuzzy rule. The last gene can have value 1, representing the positive (minority) class, or 2, representing the negative (majority) class. The maximum value for the other genes is $\#LT$. The minimum value for the other genes is 0; the values 0 represent a don't care condition which is a condition that does not appear in the rule, i.e., the corresponding linguistic variable was removed from the rule antecedent. An example of a fuzzy rule used in this work with five conditions and a consequent is showed in Figure 2.22.

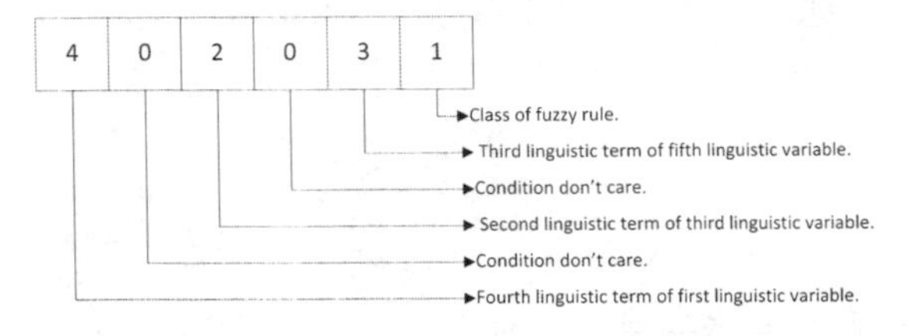

FIGURE 2.22: Fuzzy rule representation.

The MOEA used in the method optimizes two conflicting objectives: the accuracy and the interpretability of the rule. Accuracy is computed by means of the Completeness (Com) and the Consistency (Con) of each rule [102]. In this work, we consider the Com as the difference between two values: the sum of the compatibility grades of a fuzzy rule with not marked examples of the same class as the rule multiplied by the number of such examples; and the sum of the compatibility grades of the fuzzy rule with the examples of classes different from the class of the rule multiplied by the number of such examples (see Section 2.5.4 for a representation of the marked examples concept). Equation 2.4 defines the calculation of Com index for a rule with class c, denoted by R_c. The Con is defined as the ratio between two values: the number of not marked examples from class c that have a high compatibility degree with the rule and the number of examples that are not from class c and have a high compatibility degree with the Equation 2.5.

$$\begin{aligned} Com(R_c) = & \left[1 + \sum_{e=1}^{n} g(e_c \notin lm)\right] \times |\forall e_c \notin lm| \\ & - \left[1 + \sum_{e=1}^{n} g(e_{\neq c})\right] \times |\forall e_{\neq c}|, \end{aligned} \tag{2.4}$$

$$Con(R_c) = \frac{1 + |g(e_c \notin lm) \geq gm|}{1 + |g(e_{\neq c}) \geq gm|}, \tag{2.5}$$

where:

R_c : Fuzzy rule with class c.
g : Compatibility degree.
e_c : Example e is an example of the class c.
$e_{\neq c}$: Example e is not an example of the class c.
lm : Set of marked examples.
gm : Example degree marking or $markValue$.

Accuracy is an objective to be minimized and is computed as:

$$Accuracy\ (R) \ = -(Com \times Con). \tag{2.6}$$

Interpretability is measured by the number of conditions in the rule. It is an objective to be minimized and is computed as:

$$Interpretability\ (R) \ = \forall cond \neq don't\ care. \tag{2.7}$$

For the crossover operator, we use the SBX [66] which works with two parent fuzzy rules and creates two children fuzzy rules. The operator SBX is applied with a probability $ProbSBX$. Spread factor β, defined in Equation 2.8, is calculated as the ratio of the distance between the children to that of the parents and depends on a random number u. The values of each child gene are calculated according to Equations 2.9 and 2.10.

$$\beta = \begin{cases} (2u)^{\frac{1}{n+1}}, & if \quad u \leqslant 0.5 \\ \left(\frac{1}{2(1-u)}\right)^{\frac{1}{n+1}}, & if \quad u > 0.5 \end{cases}. \tag{2.8}$$

$$x_1^{child} = 0.5\left[(1+\beta)x_1 + (1-\beta)x_2\right]. \tag{2.9}$$

$$x_2^{child} = 0.5\left[(1-\beta)x_1 + (1+\beta)x_2\right]. \tag{2.10}$$

An example of application of this crossover operator on two rules with different consequent classes is shown in Figure 2.23(a). The children fuzzy rules are calculated from prechildren fuzzy rules because some values may not be integers (we consider an integer floor of the gene value), or can be lower or higher than the minimum and maximum values for each gene mentioned above, respectively. The mutation operator used is the polynomial mutation [66] (it was improved in [74]). It is based on the polynomial distribution where the variance of each value is a function of the distribution control parameter *cp*. The polynomial mutation is applied with a probability $ProbPoliMuta$. An example of this mutation in two rules is shown in Figure 2.23(b). Similar to what is done with an SBX operator, the fuzzy rules mutated are calculated from prefuzzy rules mutated because some values may not be integers (we consider an integer floor of the gene value), or can be lower or higher than the minimum and maximum values for each gene mentioned above, respectively. Finally, we consider a number of generations, $generationsNSGA\text{-}II$, as the stop condition.

2.5.3 Select and Insert the Best Rule into the Rule Base

We have to select and insert in the RB the best rule in the non-dominated fuzzy rules set obtained in the previous step. For this task, we follow the next substeps:

Parent Fuzzy Rule 1	4	0	2	0	3	1
Parent Fuzzy Rule 2	0	3	1	5	0	2
Pre Children Fuzzy Rule 1	3.5	0.3	1.9	0.6	2.7	1.1
Pre Children Fuzzy Rule 2	0.5	2.7	1.1	4.4	0.3	1.9
Children Fuzzy Rule 1	3	0	1	0	2	1
Children Fuzzy Rule 2	0	2	1	4	0	1

(a)

Fuzzy Rule	3	0	1	0	2	1
Pre Fuzzy Rule Mutated	3	2.2	1	0	2	2.1
Fuzzy Rule Mutated	3	2	1	0	2	2

(b)

FIGURE 2.23: (a) Example of application of SBX operator, random number $u = 0.3$ Spread Factor $\beta = 0.7746$. (b) Example of application of mutation operator, distribution Control Parameter $cp = 0.8$.

1. Sort the non-dominated fuzzy rules set by the number of conditions.
2. If the number of non-dominated fuzzy rules is odd, select the fuzzy rule on the middle position. If the number of non-dominated fuzzy rules is even, select a random fuzzy rule between the rules in the two middle positions.
3. Insert the selected fuzzy rule into the RB.
4. If the classification rate of the RB improves, leave the rule in the RB and continue the learning process passing to the next step (marking the examples and moving to the next run of the MOEA).
5. If the classification rate of the RB doesn't improve, remove the selected fuzzy rule from the non-dominated fuzzy rule set and from the RB. If the non-dominated fuzzy rules set is not empty, return to substep 2. When the non-dominated fuzzy rules set is empty, the classification fuzzy rule learning process finishes.

The classification rate is calculated using the classic fuzzy reasoning method described in Section 2.3.2.

2.5.4 Marked Examples Covered by the Best Rule

The marked examples are important in the accuracy objective as we showed above. The completeness and consistency of a rule are measured taking into account the degree of compatibility between the rule and the example coverage by the rule. If a selected rule is inserted into the RB all examples with a compatibility degree with the rule higher than $markValue$ are marked. In the next run of the MOEA, the recently marked examples will influence the evaluation of completeness and consistency measures. This influence will favor the selection of a rule that has high compatibility with examples that are not covered by the rule selected in the previous run of the algorithm. After that, the classification fuzzy rule learning process is repeated.

2.6 Experimental Analysis

In this section, we present the experiments developed to analyze the behavior of IRL-ID-MOEA. Table 2.1 summarizes the values of parameters used for the proposed method in the experiments. The meaning of each parameter has been presented in the preceding sections.

TABLE 2.1: Parameters used in the experiments

Parameter	Value
$\#LT$	5
$sizeNSGA\text{-}II$	100
$generationsNSGA\text{-}II$	1000
$ProbSBX$	90%
$ProbPoliMuta$	10%
$markValue$	0.3

We employ the proposed method on 22 imbalanced datasets with a high IR (> 9.0). These imbalanced datasets were extracted from KEEL repository [2] and are characterized by the number of examples (#Exam.), number of linguistic variables (#Vari.), and the IR, as shown in Table 2.2.

All experiments were run using five-fold cross validation. The next six tables show the results (for space sake, we show the results for seven imbalanced datasets) of accuracy and interpretability for IRL-ID-MOEA using 12 preprocessing techniques mentioned in Section 2.2 and without processing technique (None). The accuracy was evaluated by means of the Area Under the Curve (AUC) metric [120] which is suggested as a suitable measure in imbalanced domains. The interpretability was evaluated in terms of the number of fuzzy rules (#Rules) and conditions (#Cond.). The standard deviation is shown in brackets.

Table 2.9 shows the best results of AUC for each dataset. Based on this table we conclude that the proposed methods have better results when an oversampling or hybrid method is used than when an undersampling method is used.

TABLE 2.2: Description of imbalanced datasets

Dataset	#Exam.	#Vari.	*IR*
Abalone19	4174	8	128.87
Abalone9vs18	731	8	16.68
Ecoli0137vs26	281	7	39.15
Ecoli4	336	7	13.84
Glass016vs2	192	9	10.29
Glass016vs5	184	9	19.44
Glass2	214	9	10.39
Glass4	214	9	15.47
Glass5	214	9	22.81
Page-blocks13vs2	472	10	15.85
Shuttle0vs4	1829	9	13.87
Shuttle2vs4	129	9	20.5
Vowel0	988	13	10.1
Yeast05679vs4	528	8	9.35
Yeast1289vs7	947	8	30.56
Yeast1458vs7	693	8	22.1
Yeast1vs7	459	8	13.87
Yeast2vs4	514	8	9.08
Yeast2vs8	482	8	23.1
Yeast4	1484	8	28.41
Yeast5	1484	8	32.78
Yeast6	1484	8	39.15

TABLE 2.3: Result of the proposed method IRL-ID-MOEA for Abalone19 imbalanced dataset

Dataset	**Preprocessing**	**AUC**	**Rules**	**Conditions**
Abalone19	None	0.500 (0.000)	2.800 (1.095)	2.800 (1.095)
Abalone19	SMOTE	0.771 (0.101)	5.800 (1.924)	9.400 (3.647)
Abalone19	Borderline_SMOTE	0.676 (0.103)	6.200 (1.304)	11.400 (3.847)
Abalone19	ADASYN	0.767 (0.114)	5.400 (2.191)	9.400 (5.683)
Abalone19	Safe_Level_SMOTE	0.783 (0.106)	5.600 (1.817)	10.600 (4.393)
Abalone19	TL	0.500 (0.000)	3.200 (1.095)	3.200 (1.095)
Abalone19	OSS	0.500 (0.001)	1.800 (0.447)	2.000 (0.707)
Abalone19	NCL	0.500 (0.000)	2.400 (0.894)	2.400 (0.894)
Abalone19	SBC	0.500 (0.000)	1.000 (0.000)	1.000 (0.000)
Abalone19	SMOTE_TL	0.775 (0.102)	6.200 (1.789)	11.400 (4.037)
Abalone19	SMOTE_ENN	0.773 (0.109)	6.400 (1.140)	12.200 (3.564)

TABLE 2.4: Result of the proposed method IRL-ID-MOEA for Ecoli0137vs26 imbalanced dataset

Dataset	Preprocessing	AUC	Rules	Conditions
Ecoli0137vs26	None	0.546 (0.109)	4.400 (0.894)	4.600 (1.140)
Ecoli0137vs26	SMOTE	0.842 (0.196)	7.800 (1.483)	11.600 (2.702)
Ecoli0137vs26	Borderline_SMOTE	0.815 (0.219)	6.600 (1.517)	7.400 (1.817)
Ecoli0137vs26	ADASYN	0.809 (0.197)	6.200 (2.683)	9.400 (4.336)
Ecoli0137vs26	Safe_Level_SMOTE	0.820 (0.206)	8.000 (2.121)	12.800 (2.588)
Ecoli0137vs26	TL	0.846 (0.221)	5.200 (0.447)	5.800 (0.447)
Ecoli0137vs26	OSS	0.772 (0.082)	2.200 (0.447)	2.600 (0.548)
Ecoli0137vs26	NCL	0.796 (0.275)	5.400 (0.548)	6.000 (0.707)
Ecoli0137vs26	SBC	0.500 (0.000)	1.000 (0.000)	1.000 (0.000)
Ecoli0137vs26	SMOTE_TL	0.849 (0.201)	7.000 (1.225)	10.800 (3.114)
Ecoli0137vs26	SMOTE_ENN	0.845 (0.198)	7.400 (1.517)	9.600 (3.050)

TABLE 2.5: Result of the proposed method IRL-ID-MOEA for Glass016vs5 imbalanced dataset

Dataset	Preprocessing	AUC	Rules	Conditions
Glass016vs5	None	0.497 (0.006)	2.600 (0.894)	2.800 (1.304)
Glass016vs5	SMOTE	0.896 (0.127)	7.200 (2.049)	14.200 (6.535)
Glass016vs5	Borderline_SMOTE	0.827 (0.233)	7.200 (1.483)	12.600 (2.510)
Glass016vs5	ADASYN	0.934 (0.036)	5.600 (0.548)	9.400 (1.817)
Glass016vs5	Safe_Level_SMOTE	0.826 (0.238)	6.000 (1.581)	11.800 (2.387)
Glass016vs5	TL	0.600 (0.224)	3.000 (1.000)	3.600 (1.817)
Glass016vs5	OSS	0.713 (0.244)	3.200 (1.095)	5.800 (1.643)
Glass016vs5	NCL	0.647 (0.226)	3.200 (0.837)	3.600 (1.342)
Glass016vs5	SBC	0.500 (0.000)	1.000 (0.000)	1.000 (0.000)
Glass016vs5	SMOTE_TL	0.946 (0.027)	7.400 (0.548)	15.600 (3.578)
Glass016vs5	SMOTE_ENN	0.943 (0.045)	5.600 (1.342)	13.000 (4.848)

Considering these results, we select the SMOTE+TL method to compare the proposed method with another method for learning fuzzy rules from the imbalanced dataset in the literature. Table 2.10 shows the same results for the GA-FS-GL method [238] that balances the imbalanced dataset using SMOTE and uses a GA for jointly performing a feature selection and granularity learning for FRBC. The best values for each performance measure between the two methods are marked in bold.

From Table 2.10 we can observe that the accuracy based on AUC obtained by IRL-ID-MOEA method is higher for 17 imbalanced datasets, although the AUC average is similar for both methods. The average number of fuzzy rules obtained by IRL-ID-MOEA method is less than the average obtained by GA-FS+GL method but both methods are better on 11 imbalanced datasets.

TABLE 2.6: Result of the proposed method IRL-ID-MOEA for Shuttle0vs4 imbalanced dataset

Dataset	Preprocessing	AUC	Rules	Conditions
Shuttle0vs4	None	0.992 (0.011)	3.000 (0.000)	3.000 (0.000)
Shuttle0vs4	SMOTE	0.992 (0.011)	3.600 (0.548)	5.000 (1.225)
Shuttle0vs4	Borderline_SMOTE	0.992 (0.011)	3.000 (0.000)	3.000 (0.000)
Shuttle0vs4	ADASYN	0.986 (0.018)	2.800 (1.095)	3.400 (1.673)
Shuttle0vs4	Safe_Level_SMOTE	0.992 (0.011)	3.400 (0.548)	3.600 (0.548)
Shuttle0vs4	TL	0.992 (0.011)	3.000 (0.000)	3.000 (0.000)
Shuttle0vs4	OSS	0.959 (0.047)	2.800 (0.447)	3.200 (0.837)
Shuttle0vs4	NCL	0.992 (0.011)	3.400 (0.548)	3.400 (0.548)
Shuttle0vs4	SBC	0.496 (0.009)	1.600 (0.548)	1.600 (0.548)
Shuttle0vs4	SMOTE_TL	0.992 (0.011)	4.800 (1.095)	6.400 (2.074)
Shuttle0vs4	SMOTE_ENN	0.992 (0.011)	4.600 (1.517)	6.200 (2.490)

TABLE 2.7: Result of the proposed method IRL-ID-MOEA for Vowel0 imbalanced dataset

Dataset	Preprocessing	AUC	Rules	Conditions
Vowel0	None	0.868 (0.078)	7.000 (1.581)	11.000 (3.240)
Vowel0	SMOTE	0.934 (0.039)	9.000 (0.000)	16.800 (0.837)
Vowel0	Borderline_SMOTE	0.917 (0.012)	7.000 (3.240)	12.000 (7.348)
Vowel0	ADASYN	0.957 (0.036)	9.600 (2.966)	21.400 (11.261)
Vowel0	Safe_Level_SMOTE	0.964 (0.009)	11.000 (1.000)	23.600 (3.847)
Vowel0	TL	0.838 (0.080)	6.200 (1.789)	9.200 (4.764)
Vowel0	OSS	0.898 (0.071)	5.800 (1.789)	10.200 (4.764)
Vowel0	NCL	0.863 (0.070)	6.000 (1.581)	8.800 (3.962)
Vowel0	SBC	0.751 (0.230)	4.400 (2.510)	7.600 (5.899)
Vowel0	SMOTE_TL	0.948 (0.037)	9.600 (2.302)	19.400 (6.066)
Vowel0	SMOTE_ENN	0.936 (0.056)	9.800 (1.095)	20.000 (2.449)

The biggest difference between the two methods can be seen in the number of conditions, where IRL-ID-MOEA obtains lower values for 17 datasets, and the average value is less than half the average obtained by GA-FS+GL method.

In order to give statistical support to the obtained results, we carried out a Wilcoxon signed-rank test [242] [166] to compare both methods. The test values are shown in Table 2.11. In the test, the p-value for the statistical distribution is computed and if it is below a specified level of significance α the null hypothesis of equality of means can be rejected meaning that there exists a significant difference between both methods. The Wilcoxon signed-rank test shows that there exists a significant difference in the AUC and the number of conditions with $\alpha = 0.05$.

TABLE 2.8: Result of proposed method IRL-ID-MOEA for Yeast1458vs7 imbalanced dataset

Dataset	**Preprocessing**	**AUC**	**Rules**	**Conditions**
Yeast1458vs7	None	0.498 (0.002)	1.800 (0.837)	1.800 (0.837)
Yeast1458vs7	SMOTE	0.660 (0.056)	5.200 (2.387)	10.200 (5.263)
Yeast1458vs7	Borderline_SMOTE	0.620 (0.069)	7.000 (3.162)	15.000 (7.176)
Yeast1458vs7	ADASYN	0.635 (0.037)	6.000 (2.000)	11.800 (5.805)
Yeast1458vs7	Safe_Level_SMOTE	0.632 (0.050)	7.200 (2.168)	15.200 (5.630)
Yeast1458vs7	TL	0.497 (0.002)	1.200 (0.447)	2.200 (2.683)
Yeast1458vs7	OSS	0.497 (0.002)	1.000 (0.000)	1.000 (0.000)
Yeast1458vs7	NCL	0.496 (0.000)	1.000 (0.000)	1.000 (0.000)
Yeast1458vs7	SBC	0.500 (0.000)	1.000 (0.000)	1.000 (0.000)
Yeast1458vs7	SMOTE_TL	0.627 (0.031)	7.000 (3.317)	14.600 (7.893)
Yeast1458vs7	SMOTE_ENN	0.641 (0.083)	8.200 (2.864)	18.800 (8.167)

TABLE 2.9: Best results of AUC of the proposed method IRL-ID-MOEA for each imbalanced dataset

Dataset	**Preprocessing**	**AUC**	**Rules**	**Conditions**
Abalone19	Safe_Level_SMOTE	0.783 (0.106)	5.600 (1.817)	10.600 (4.393)
Abalone9vs18	Borderline_SMOTE	0.772 (0.127)	4.800 (1.095)	8.000 (2.449)
Ecoli0137vs26	SMOTE_TL	0.849 (0.201)	7.000 (1.225)	10.800 (3.114)
Ecoli4	SMOTE_ENN	0.957 (0.030)	6.800 (1.304)	12.400 (3.435)
Glass016vs2	Safe_Level_SMOTE	0.706 (0.068)	4.600 (1.140)	6.600 (2.966)
Glass016vs5	SMOTE_TL	0.946 (0.027)	7.400 (0.548)	15.600 (3.578)
Glass2	Safe_Level_SMOTE	0.731 (0.059)	5.200 (1.483)	9.000 (3.742)
Glass4	Borderline_SMOTE	0.942 (0.070)	6.800 (1.643)	13.600 (3.050)
Glass5	SMOTE_TL	0.889 (0.106)	8.200 (3.347)	16.200 (6.058)
Page-blocks13vs2	Borderline_SMOTE	0.980 (0.009)	5.400 (1.140)	9.600 (1.949)
Shuttle0vs4	SMOTE_TL	0.992 (0.011)	4.800 (1.095)	6.400 (2.074)
Shuttle2vs4	NCL	0.992 (0.011)	2.200 (0.447)	2.200 (0.447)
Vowel0	ADASYN	0.957 (0.036)	9.600 (2.966)	21.400 (11.261)
Yeast05679vs4	SMOTE	0.831 (0.035)	6.600 (1.517)	13.400 (3.362)
Yeast1289vs7	ADASYN	0.748 (0.055)	6.600 (1.817)	15.000 (3.742)
Yeast1458vs7	SMOTE	0.660 (0.056)	5.200 (2.387)	10.200 (5.263)
Yeast1vs7	ADASYN	0.659 (0.057)	8.600 (3.130)	17.600 (8.989)
Yeast2vs4	SMOTE_ENN	0.910 (0.029)	5.800 (1.304)	11.400 (2.793)
Yeast2vs8	SMOTE_ENN	0.791 (0.106)	2.800 (0.447)	4.200 (1.643)
Yeast4	Borderline_SMOTE	0.856 (0.030)	5.800 (1.483)	12.000 (2.739)
Yeast5	SMOTE_ENN	0.959 (0.016)	4.200 (0.447)	7.200 (1.924)
Yeast6	Borderline_SMOTE	0.914 (0.069)	7.400 (0.548)	12.600 (2.191)

TABLE 2.10: Average values of the accuracy and interpretability for GA-FS+GL and IRL-ID-MOEA

Dataset	GA-FS+GL			IRL-ID-MOEA		
	AUC	#Rules	#Cond.	AUC	#Rules	#Cond.
Abalone19	0.699	7.400	14.800	**0.775**	**5.800**	**9.400**
Abalone9-18	0.628	**4.400**	12.320	**0.714**	4.800	**8.000**
Ecoli0137vs26	0.790	**4.200**	**10.080**	**0.849**	7.000	10.800
Ecoli4	0.860	58.200	162.960	**0.927**	**6.800**	**12.400**
Glass016vs2	0.626	**3.600**	10.800	**0.633**	4.600	**6.600**
Glass016vs5	0.867	**4.000**	**12.000**	**0.946**	7.200	12.600
Glass2	**0.729**	8.200	24.600	0.670	**5.200**	**9.000**
Glass4	0.853	8.400	26.880	**0.913**	**6.800**	**13.600**
Glass5	0.755	**5.200**	**15.600**	**0.889**	8.200	16.200
Page-Blocks13vs4	0.937	6.600	21.120	**0.945**	**5.400**	**9.600**
shuttle0vs4	**0.999**	14.800	32.560	0.992	**3.000**	**3.000**
shuttle2vs4	**0.992**	4.200	12.600	0.892	**2.200**	**2.200**
Vowel0	0.928	**4.400**	22.880	0.948	9.600	**21.400**
Yeast05679vs4	0.793	9.400	20.680	**0.812**	**6.600**	**13.400**
Yeast1289vs7	**0.750**	13.800	41.400	0.717	**6.600**	**15.000**
Yeast1458vs7	**0.641**	**3.800**	**8.360**	0.627	5.200	10.200
yeast1vs7	**0.754**	**6.800**	19.040	0.632	8.600	**17.600**
Yeast2vs4	0.884	**2.200**	**6.600**	**0.905**	4.400	8.200
Yeast2vs8	0.710	5.000	27.000	**0.786**	**2.800**	**4.200**
Yeast4	0.835	**6.400**	17.920	**0.851**	7.200	**13.200**
Yeast5	0.935	4.800	14.400	**0.951**	**4.600**	**7.600**
Yeast6	0.870	**5.000**	14.000	**0.887**	7.400	**12.600**
Average	0.811	8.673	24.936	**0.830**	**5.909**	**10.764**

TABLE 2.11: Results of Wilcoxon signed-rank test between GA-FS+GL and IRL-ID-MOEA

	AUC	#Rules	#Cond.
p-value	0.069	0.758	0.001

2.7 Final Remarks

In this chapter we described a method that combines preprocessing methods to balance the number of examples, the IRL approach, and a MOEA for learning fuzzy classification rules from imbalanced datasets. The proposed method, called IRL-ID-MOEA, integrates techniques of two levels for dealing with the imbalanced problems: preprocessing methods, that are techniques that focus on the data level, and a MOEA with don't care conditions, a technique that focuses on the feature selection level. In the work developed here, ten different algorithms to balance the number of examples in the classes were used in the first phase of the method, the preprocessing phase. In the second phase, a MOEA, specifically the NSGA-II, is applied to learn a set of non-dominated classification fuzzy rules. Two objectives are optimized by the MOEA: accuracy and interpretability. As a means to analyze the performance of our method, the IRL-ID-MOEA proposed method (using SMOTE+TL technique) was compared to the GA-FS+GL method (using SMOTE technique) in order to demonstrate its good performance. The experimental analysis demonstrated that the combination used in the work described here is a promising approach for learning FRBCSs on imbalanced datasets with a good balance between accuracy and interpretability.

In future research we plan to use other balance methods and other MOEAs with the possibility for improving accuracy and/or interpretability of the FR-BCS. Moreover, we want to analyze different ways to select the best rule in a non-dominated fuzzy rules set and to consider other interpretability objectives in fuzzy classification rule learning or fuzzy sets tuning process.

Chapter 3

Hybrid Multi-Objective Evolutionary Algorithms with Collective Intelligence

Daniel Cinalli

Instituto de Computação, Universidade Federal Fluminense

Luis Martí

Instituto de Computação, Universidade Federal Fluminense

Nayat Sanchez-Pi

Instituto de Matemática e Estatística, Universidade do Estado do Rio de Janeiro

Ana Cristina Bicharra García

Instituto de Computação, Universidade Federal Fluminense

3.1 Introduction

Many real-world optimization problems can be formulated as multi-objective optimization problems (MOPs), in which two or more objective functions must be simultaneously optimized [82]. In the general case, the solution of a MOP is not a single point that optimizes all the objectives at the same time, but, instead, a set of points that represent different trade-offs between the objectives known as Pareto-optimal set. In such a situation, a decision maker (DM) must select which of those solutions are the ones to be carried out in practice according to some *a priori* high-level preferences.

Optimization problems and, hence, MOPs are NP-complete [10]. Therefore, deterministic search techniques are usually unsuitable to handle the complexity of this task. Metaheuristic and stochastic approaches are a viable alternative to handle MOPs. Multi-objective evolutionary algorithms (MOEAs) [48] are one example of such methods. MOEAs can —or perhaps, must— take advantage of the DM preferences to drive the search process by focusing on solutions of interest, instead of all possible solutions.

Expressing a preference from a unique decision maker in the optimization process may raise some concerns regarding unilateral choice bias. In contrast, a global outcome built on the aggregation of vast and diverse masses of individual DM intelligence would be a helpful input parameter not only to guide the search, but also to explore a wider diversity of answers and enhance the partial results through multi-user interaction.

The present work proposes two new collective preference-based interactive MOEAs based on the non-dominated sorting genetic algorithm II [69] (CI-NSGA-II) and the S-metric selection algorithm [22] (CI-SMS-EMOA), respectively. These algorithms combine two distinct fields not connected before as it aggregates consistent preferences from a collective intelligence (COIN) [229] environment to the optimization process of MOEAs. They use people's heterogeneity and common sense to support one or many experienced DMs in the search for relevant regions in Pareto-optimal set. Built upon the subjectivity of the crowds and human cognition, the intelligence of participatory actions addresses dynamic collective reference points to overcome MOPs' difficulties and guide the exploration of preferred solutions.

This method improves the quality of the obtained Pareto frontier approximation. It has an advantage over traditional iterative approaches because their results are driven not by one DM, but a group of people that delimits their collective area of interest in the objective space. The new algorithms produce better solutions in the sense that they iteratively refine the search parameters and generate points more appropriated for DM's choice.

The rest of this chapter is organized as follows. Section 3.2 covers some required formal definitions of multi-objective optimization and collective intelligence field. Section 3.3 and 3.4 outline the usage of preferences and collective intelligence in MOEAs, respectively. Section 3.5 presents the new algorithms CI-NSGA-II and CI-SMS-EMOA based on interactive collective intelligence techniques. Some results from benchmark problems and a resource placement case study are analyzed in Section 3.6. Finally, in Section 3.7, conclusive remarks and future work directions are put forward.

3.2 Foundations

A MOP can be presented as the minimization of the functions $F(\boldsymbol{x}) = \{f_1(\boldsymbol{x}), \ldots, f_k(\boldsymbol{x})\}$ subject to a set of constraints $g_i(\boldsymbol{x}) \leq 0, i = 1, \ldots, m$; $h_j(\boldsymbol{x}) = 0, j = 1, \ldots, p$; where $\boldsymbol{x} = \langle x_1, \ldots, x_n \rangle \in \Omega$ is an n-dimensional vector of decision variable. Thus, a MOP consists of k objectives, $m + p$ constraints, n decision variables, and an evaluation function $\boldsymbol{F} : \Omega \rightarrow \mathcal{Z}$ that maps from the vector $\boldsymbol{x}$ to output vectors $\boldsymbol{a} = \langle a_1, \ldots, a_k \rangle$. The solution to this problem can be expressed by relying on the Pareto dominance relationship. A $\boldsymbol{x}$ is said to dominate $\boldsymbol{v}$ (denoted as $\boldsymbol{x} \prec \boldsymbol{v}$) iff $\forall i \in \{1, \ldots, n\}$, $x_i \leq v_i \wedge \exists i \in \{1, \ldots, n\}$ such that $x_i < v_i$. Likewise, $\boldsymbol{x}$ weakly dominates $\boldsymbol{v}$ ($\boldsymbol{x} \preccurlyeq \boldsymbol{v}$) iff $\forall i \in \{1, \ldots, n\}$, $x_i \leq v_i$. The strictly dominant $\boldsymbol{x} \prec\!\!\prec \boldsymbol{v}$ stands only iff $\forall i \in \{1, \ldots, n\}$, $x_i < v_i$. The decision variables can also be incomparable ($\boldsymbol{x} \parallel \boldsymbol{v}$) when $\neg(\boldsymbol{x} \preccurlyeq \boldsymbol{v}) \wedge \neg(\boldsymbol{v} \preccurlyeq \boldsymbol{x})$.

A solution $\boldsymbol{x} \in \Omega$ is Pareto optimal if there does not exist another solution $\boldsymbol{x}' \in \Omega$ such that $F(\boldsymbol{x}') \prec F(\boldsymbol{x})\}$. The Pareto-optimal set, P_S, is defined as $P_S = \{\boldsymbol{x} \in \Omega, \nexists \boldsymbol{x}' \in \Omega$ such that $F(\boldsymbol{x}') \prec F(\boldsymbol{x})\}$. Similarly, the codomain of the set is known as the Pareto-optimal front, P_F [49].

3.2.1 Evolutionary Multi-Objective Optimization

Evolutionary algorithms (EAs) are a successful alternative for multi-objective optimization because of its search technique that relies on finite population of candidate solutions and the generation of many possible answers in every single run. Multi-objective EAs (MOEAs) follow the common concepts of EAs and the underlying idea is to identify the set of individuals non-dominated by the rest of the population in every generation. Their typical components are a space of individuals I to be considered as candidate solutions, a problem-specific fitness function of individuals $F : I \rightarrow \mathbb{R}$ to measure if a certain solution satisfies the objective functions, and some operators in charge of tasks ranging from reproduction of the individuals (crossover and mutation) to pushing selection and competition pressure on them. After running a MOEA, the final population detains an approximation set (S) of all non-dominated solutions with finite size that can be an appropriate representation of P_S.

Many techniques are used to evaluate the quality of MOEA solutions. One of them is the hypervolume or S-metric indicator [84]. It calculates the volume of the union of hypercubes a_i defined by a non-dominated point m_i and a reference point x_{ref} defined as:

$$\begin{aligned} S(M) &= \Lambda(\{\textstyle\bigcup_i a_i | m_i \in M\}) \\ &= \Lambda(\textstyle\bigcup_{m \in M}\{x | m \prec x \prec x_{ref}\}) \, . \end{aligned} \tag{3.1}$$

It is a quantitative metric that computes the region space covered by all non-dominated points. This performance indicator can be used independently to evaluate the efficiency of different multi-objective algorithms and does not require knowledge of the true Pareto-optimal front beforehand.

Pareto-optimal front coverage indicator, $D_{S \to P_F}$, is a proximity indicator [29] that defines the distance between an achieved approximation set S and their closest counterpart in the current Pareto-optimal front:

$$D_{S \to P_F}(S) = \frac{1}{|S|} \sum_{\boldsymbol{x} \in S} \min_{\boldsymbol{x}' \in P_S} \{d(\boldsymbol{x}, \boldsymbol{x}')\}, \tag{3.2}$$

where d is the Euclidean distance between two points. If the Pareto-optimal front is continuous, a correct formulation of this indicator calls for a line integration over S. Small values of $D_{S \to P_F}$ indicate proximity to the Pareto-optimal front. Its main drawback is the obligation to previously know the true front, which may be unfeasible for real applications.

The variance (σ^2) is a statistical measure that expresses the dispersion of data. Instead of a good spread of solutions along P_F, the method proposed in this work wants to obtain subsets of solutions close to the collective reference point. In this context, a small variance means the individuals from the sample $Y = \{\boldsymbol{y}_1, \ldots, \boldsymbol{y}_N\}$ are clustered closely around the population mean (μ). A low dispersion for a group of preferred points in P_F denotes a better efficiency of the approach tested.

$$\sigma^2 = \frac{1}{N} \sum_{i=1}^{N} (\boldsymbol{y_i} - \mu)^2. \tag{3.3}$$

In cases with more than one collective reference point ($\boldsymbol{z}^j$), the points are clustered based on the closest distance to one of the reference points: $C_j = \{\boldsymbol{a} \in \mathbb{R}^k : \|\boldsymbol{a} - \boldsymbol{z}^j\| \leq \|\boldsymbol{a} - \boldsymbol{z}^i\|, \forall i\}$. Cluster C_j consists of all points for which $\boldsymbol{z}^j$ is the closest. The variance is calculated to each cluster separately.

3.2.2 Collective Intelligence

Since the beginning of 2000, the development of social network technologies and interactive online systems has promoted a broader understanding of the "intelligence" concept. A new phenomenon appeared based not only on the cognition of one individual, but also placed on a network of relationships with other people and the external world. The field known as collective intelligence (COIN) [165] is defined as the self-organized group intelligence arisen from participatory and collaboration actions of many individuals. Shared tasks are handled by singular contributions in such a manner that their aggregation process creates better results and solves more problems than each particular contribution separately [229]. This phenomenon develops a *sui generis* intelligence. It raises a global experience of collective attitudes without centralized control, bigger than its isolated pieces and subproduct of their combination.

COIN involves groups of individuals collaborating to create synergy and augment the intellectual processes of human beings. A decision-making process over the Internet has to manage users' interactions. It must get valuable knowledge concealed or dispersed in the group, even when the participants are not specialized in the subject. This environment includes large and heterogeneous audiences that are mostly independent among each other. Therefore, the problem must be decomposed in tasks that sustain diversity and transient members' attendances to align the interest of crowds.

Handling collective intelligence means the combination of those tools and methods in a dynamic space of production to achieve an objective. As the purpose of this study is the enhancement of MOEAs through the use of the collective, interactive genetic algorithms (IGA) are an appropriate technique to support this goal. IGA incorporates the evaluation of users on the candidates of evolutionary algorithms to solve problems whose optimization objectives are complex to be defined with exact functions [227]. Users' subjectivities are employed as fitness values to drive the search throughout the evolution process.

3.3 Preferences and Interactive Methods

The current state-of-the-art MOEAs are capable of obtaining reliable approximations of the Pareto-optimal front. Some MOEAs like SMS-EMOA [22], MO-CMA-ES [122], NSGA-II [69], or SPEA2 [268] exploit a set of solutions in parallel and supply a starting point to new developments in the field.

However, the optimal frontier might be extremely large or possibly infinite and the DMs still must identify a final answer to their demands from this trade-off set. The challenge is no longer just to obtain a diversity of answers in the entire high-dimensional P_F, but also retrieve the expected solutions aligned to consistent preferences of the DMs. In most of the cases, their preferences are determined in the objective space.

Preferences are user-defined parameters and denote values or subjective impressions regarding the trade-offs points. It transforms qualitative feelings into quantitative values to bias the search during the optimization phase and restrict the objective space. In this sense, a reliable preference vector improves the trade-off answers obtained. Local preferences can be expressed as a set of constraints, as a vector of weights or a lexicographic sorting of objectives, or as trade-off information or reference points for the search, among other representations.

Wierzbicki's reference point approach [241] concentrates the search of non-dominated solutions near the selected point. It is based on the achievement scalarizing function that uses a reference point to capture the desired values of the objective functions. Let $\boldsymbol{z}^0$ be a reference point for an n-objective optimization problem of minimizing $F(\boldsymbol{x}) = \{f_1(\boldsymbol{x}), ..., f_k(\boldsymbol{x})\}$. The reference point scalarizing function can be stated as follows:

$$\sigma\left(\boldsymbol{z}, \boldsymbol{z}^0, \boldsymbol{\lambda}, \rho\right) = \max_{i=1,\ldots,k} \{\lambda_i(z_i - z_i^0)\} + \rho \sum_{i=1}^{k} \lambda_i \left(z_i - z_i^0\right), \quad (3.4)$$

where $\boldsymbol{z} \in \mathcal{Z}$ is one objective vector, $\boldsymbol{z}^0 = \langle z_1^0, ..., z_k^0 \rangle$ is a reference point vector, σ is a mapping from $\mathbb{R}^k$ onto $\mathbb{R}$, $\boldsymbol{\lambda} = \langle \lambda_1, ..., \lambda_k \rangle$ is a scaling coefficient vector, and ρ is an arbitrary small positive number. Therefore, the achievement problem can be rebuilt as: min $\sigma\left(\boldsymbol{z}, \boldsymbol{z}^0, \boldsymbol{\lambda}, \rho\right)$.

Since the 1980s, there have been several works on interactive multi-objective methods using reference points and reference directions as preferences. Those approaches were applied mainly in the classical multi-objective programming field. But in the last 15 years they have also emerged in the MOEA area.

Deb et al. [71] proposed a reference-point-based NSGA-II procedure (R-NSGA-II) to find a set of solutions in the neighborhood of the corresponding Pareto optimal. They extend the number of reference points to cover more than one region in a Pareto front and reach more significant solutions. Deb and Kumar [67] modified the NSGA-II crowding operator by the light beam search to incorporate *a priori* preferences and produce a set of solutions in the region of interest. Another variation of the original NSGA-II, the RD-NSGA-II [68] let the user supply one or more reference directions to project efficient solutions on the Pareto-optimal frontier. The principle is the application of multiple achievement scalarizing functions (σ) to generate non-dominated fronts. Ben Said et al. [21] presented a new dominance relation, named reference solution-based dominance (r-dominance), that replaces the Pareto dominance and favors solutions near the reference point indicated. Pfeiffer and others [198] rank the solutions according to their Euclidean distances to each reference point. Those points close to the reference points are assigned with the lowest crowding distance, which favors its selection by the algorithm. The negotiation support system called W-NSS-GPA [19] takes the reference points and the decision makers' hierarchy levels as weights to calculate an aggregation point of all preferences.

Interactive genetic algorithms were successfully applied to get feedbacks of transitional results throughout the evolution process. MOEAs can handle intermediate non-dominated solutions to the decision maker and improve the search with a reference point or fitness function adjustments. Deb et al. [70] proposed an interactive MOEA approach termed PI-EMO. This technique changes progressively the value function after every few generations to lead the optimization on more preferred solutions. After t generations, the user analyzes η (≥ 2) well-sparse non-dominated solutions and provides a complete or partial ranking of preference information about them. Thiele et al. [232] introduced a preference-based interactive algorithm (PBEA) that adapts the fitness evaluation with an achievement scalarizing function to guarantee an accurate approximation of the desired area in Pareto-optimal front. The algorithm IGAMII [8] applies fuzzy logic to simulate the human decision maker and relieve the constant interaction during the evolution. In BC-EMO [181], the Support Vector Ranking algorithm is used to learn an approximation of the DM utility function. iMOEA/D [100], an interactive version of the decomposition based multi-objective evolutionary algorithm, asks the DMs to analyze some current solutions and use their feedbacks to renew the preferred weight region in the following optimization.

3.4 Collective Intelligence for MOEAs

MOEA techniques are classified by its articulation of preferences: *a priori*; interactive (progressive); *a posteriori*. While some initiatives develop a partial order of preferences based on *a priori* reference points to give a stronger selection pressure among Pareto-equivalent solutions, other progressive methods combine simultaneously the preferences information and the search for solutions. But very few MOEA algorithms consider more than one user for reference point selection or evolutionary interaction. They neglect a collective scenario where many users could actively interact and take part in the decision process throughout the optimization.

Meta-heuristics approaches, like MOEAs, may confront difficulties with complex scenarios of high-dimension and large problem space. The potential number of objectives necessary to describe the environment or the incapacity to comprehend and map all the variables to a correct fitness function can prevent a solution in a reasonable time and quality. On the other hand, human beings are used to multi-objective situations in their everyday lives. Those complex scenarios that are hard for a computer might be easier or natural for a human's mind. People are able to improve the multi-objective algorithms with cognitive and subjective evaluation to find better solutions.

COIN is a different level of abstraction and can be a special contribution to make MOEAs go beyond their reach. Human characteristics such as perception, strategy, weighting factors, agility, among other subjectivities, might be introduced into the algorithm to generate a better pool of answers and enhance the optimization process. A group of people can understand a conflicting situation involving multiples objectives and may use their collective intelligence to trump an expert's abilities. The wisdom resulting from the diversity of many individuals is able to discover creative resolutions.

This work proposes the integration of collective preferences to the optimization process of MOEAs. The main idea is to apprehend people's heterogeneity and common sense to support the DM's search towards relevant regions in Pareto-optimal set. It means, in other words, the hybrid algorithms discover a subset of optimal points that is close to the collective preferences.

The method indicated here has advantages over other strategies which handle preferences in MOEA. First of all, a collective reference point produced by the interaction and aggregation of multiple opinions may provide a more accurate reference point than designed by only one DM (unilateral). A unique decision maker carries the risk of having mistaken guidelines or poor quality in terms of search parameter. Conversely, the synergy of actions and the heterogeneity inside collective environments develop creative resolutions based on the subjectivity and cognition of the crowds.

Furthermore, the new algorithms (Section 3.5) choose the reference points interactively. The references are not defined *a priori*, like the R-NSGA-II from Deb [71], nor indicated by the DM as the middle point in the Light Beam approach [67] . Rather, all the references are discovered online with the support of a genuine collective intelligence of many users.

3.5 Algorithms

This section presents the new algorithms. They are extensions of the classical MOEAs: NSGA-II [69] and SMS-EMOA [22]. The main changes on the original methods are the incorporation of COIN into the selection procedure; the transformation of the continuous evolutionary process into an interactive one; and the adoption of reference points to drive the search towards relevant regions in Pareto-optimal front.

3.5.1 CI-NSGA-II

One of the new algorithms is a variation of NSGA-II [69]. The NSGA-II is a non-domination based genetic algorithm for multi-objective optimization. It adopts two main concepts: a density information for diversity and a fast non-dominated sorting in the population. The crowding distance uses the size of the largest cuboid enclosing two neighboring solutions to estimate the density of points in the front. Solutions with higher values of this measure are preferred rather than those in a more crowded region (smaller values) because they are better contributors to a uniformly spread-out Pareto-optimal front. The non-dominated sorting places each individual into a specific front such that the first front τ_1 is a non-dominant set, the second front τ_2 is dominated only by the individuals in τ_1, and so on. Each solution inside the front τ_n receives a rank equal to its non-domination level n.

The selection operator uses the rank (i_{rank}) and crowding distance (i_{dist}) in a binary tournament. The partial order $\prec_c$ between two individuals i and j, for example, prefers the minor domination rank if they are from different fronts or, otherwise, the one with higher values of crowding distance. Then, crossover and mutation are applied to generate an offspring population.

$$i \prec_c j := i_{rank} < j_{rank} \vee (i_{rank} = j_{rank} \wedge i_{dist} > j_{dist}) \tag{3.5}$$

In algorithm 1, the new algorithm CI-NSGA-II converts the original NSGA-II into an interactive process. The subroutine *CollectivePairWise()* suspends the evolution progress and submits some individuals from population to the users' pairwise evaluation. Put differently, users select the best candidate between two or more individuals during the optimization process. Enhanced by collective subjectivity and cognition, the successive stages of evolution are improved via group's preferences in a direct crowdsourcing fashion.

Algorithm 1 The collective intelligence NSGA-II

1: $generation \leftarrow numgeneration$
2: $block \leftarrow subsetgeneration$\{iteration interval\}
3: **while** $i < generation$ **do**
4:     **while** $block$ **do**
5:         $offspring \leftarrow$ **Tournament**(pop)
6:         $offspring \leftarrow$ **Crossover**$(offspring)$
7:         $offspring \leftarrow$ **Mutation**$(offspring)$
8:         $pop \leftarrow$ **COIN Selection**$(offspring)$
9:         $i++$
10:     **end while**
11:     $comparisons \leftarrow$ **CollectivePairWise**$(front)$
12:     $\Theta \leftarrow$ **ExpectationMaximization**$(comparisons)$
13:     $pop \leftarrow$ **ReferencePoint Distance**(pop, Θ)
14: **end while**

Assuming the Central Limit Theorem [104], the inputs have a distribution that is approximately Gaussian. Therefore, after each collective interaction, the subroutine *ExpectationMaximization()* gets all the comparisons and calculates the similarity of answers. The expectation maximization approach creates online reference points (Θ) for search optimization. A Gaussian Mixture model [14] is used to emulate the evaluation landscape of all participants' choices. Figure 3.1 shows an example of three online reference points and the Gaussian distribution of their points from the well-known ZDT1 test suite [262].

The procedure *ReferencePointDistance()* calculates the minimum distance from each point in the population to the nearest collective reference points in Θ. This way, the point near the reference point is favored and stored in the new population.

The CI-NSGA-II develops a partial order similar to the NSGA-II procedure, but replaces the crowding distance operator by the distance to collective reference points (i_{ref}).

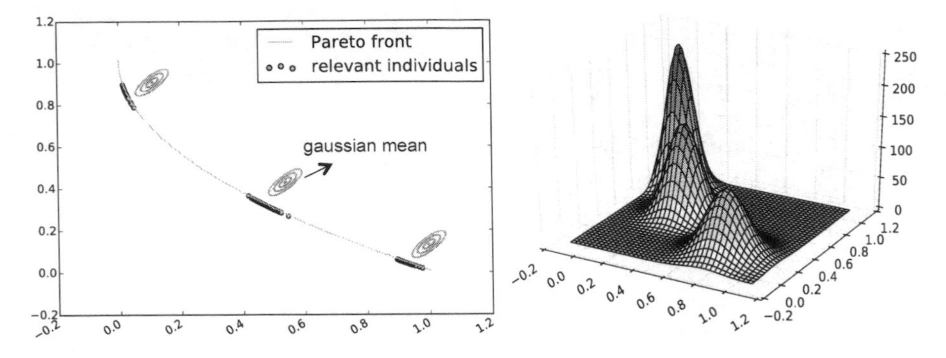

FIGURE 3.1: Three online collective reference points and their Gaussian distribution applied to the ZDT1 test.

$$i \prec_c j := i_{rank} < j_{rank} \vee (i_{rank} = j_{rank} \wedge i_{ref} < j_{ref}). \tag{3.6}$$

This algorithm performs the *COIN Selection()* operation based on the new partial order. Like NSGA-II, individuals with minor domination rank are preferred. However, if they belong to the same front, the one with the closest reference point distance is used instead.

3.5.2 CI-SMS-EMOA

The SMS-EMOA [22] is a steady-state algorithm that applies the non-dominated sorting as a ranking criterion and the hypervolume performance measure (S) as a selection operator.

After the non-domination ranking, the next step is to update the last front population, P_{worst}. It replaces the member with the minimum contribution to P_{worst} hypervolume by a new individual that increases the hypervolume covered by the population.

In algorithm 2, the new algorithm CI-SMS-EMOA converts the original SMS-EMOA into an interactive process. The subroutines *CollectivePairWise()* and *ExpectationMaximization()* work in the same way as the previous algorithm, CI-NSGA-II.

In the *COIN Selection()* operation, individuals with minor domination rank (i_{rank}) are preferred. If they belong to the same front, the one with the maximum contribution to the hypervolume of the set and the closest reference point distance (i_{ref}) is selected.

The procedure *Hype-RefPoint Distance()* gets the hypervolume contribution (S) and calculates the minimum distance from each solution in the population to the nearest collective reference points in Θ. This way, the solution with high hypervolume values and short reference point distance is favored and stored in the new population.

Algorithm 2 The collective intelligence SMS-EMOA

```
1: generation ← numgeneration
2: block ← subsetgeneration{iteration interval}
3: while i < generation do
4:   while block do
5:     offspring ← Tournament(pop)
6:     offspring ← Crossover(offspring)
7:     offspring ← Mutation(offspring)
8:     pop ← COIN Selection(offspring)
9:     i++
10:  end while
11:  comparisons ← CollectivePairWise(front)
12:  Θ ← ExpectationMaximization(comparisons)
13:  pop ← Hyper-RefPoint Distance(pop, Θ, S)
14: end while
```

3.6 Experimental Results

This section presents some results of CI-NSGA-II and CI-SMS-EMOA. The multi-objective test problems ZDT [262] and DTLZ [73] have a known optimal front and can be used to benchmark the outcome of the algorithms. A real-world case is formally introduced afterwards and submitted to a COIN experiment.

3.6.1 Multi-Objective Test Problems

ZDTs and DTLZs are a set of well-established scalable multi-objective test problems. Extensively used in MOEA studies, these benchmark problems were selected to analyze the behavior of the proposed collective intelligence MOEAs in the first moment. Each of these test functions knows *a priori* the exact shape and location of Pareto-optimal front. Their features cover different classes of MOPs: convex Pareto-optimal front, non-convex, non-contiguous convex parts, and multimodal; and in the case of DTLZ are scalable to more than two objectives ($M > 2$). For those reasons, the test problems submit the new algorithms to distinct optimization difficulties and provide a broader comprehension of its working principles.

The experiment emulates the collectivity by developing some virtual DMs (robots). Each robot has to vote between two individuals from the approximation set. The robots have a predefined point in the objective space which will be used to bias the votes. They choose an individual according to the closest distance between its predefined point and each of the two candidates. Figure 3.2 illustrates candidates $c1$ and $c2$ with their respective distances ($d1$

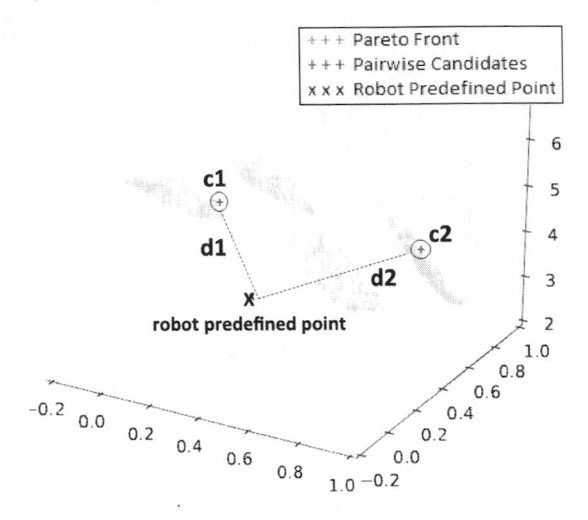

FIGURE 3.2: The robot's predefined point will choose candidate $c1$ because the distance $d1 < d2$.

and $d2$) to the predefined point x. As $d1 < d2$, the robot would vote on $c1$. The robots provide a distribution of preferences that will be used to discover online reference points

Another advantage is that one robot can perform multiple pairwise comparisons in one iteration. In the case of test problems, ZDTs and DTLZs of the robots create the collective reference points with a better reasoning than simply random choice. It is important to notice that the collective reference point is built on the similarity of answers after the Gaussian Mixture model and cannot be confused with the robots' predefined points.

The front coverage ($D_{S \rightarrow P_F}$) and the variance (σ^2) indicators were used to measure the quality of the results. The hypervolume was not employed because their values depend on the spread of solutions in the whole Pareto front and, on contrary, the proposed algorithms aim to obtain subsets of solutions close to the collective reference points.

Based on the best run of CI-NSGA-II in terms of Pareto-optimal front coverage indicator ($D_{S \rightarrow P_F}$) and variance (σ^2), Figure 3.3 shows the relevant regions found in Pareto front to the DTLZ2 and DTLZ7 problems. The quantity of online reference points is directly related to the number of k clusters in the Gaussian Mixture model. In cases where k is not previously defined, the experiment uses the X-means approach [197] to learn k from the data. This algorithm searches different values of k and scores each clustering model using the Bayesian Information Criterion (BIC): $BIC(M_j) = \iota_j(D) - (p_j/2) \log R$, where D is the dataset, M_j are models corresponding to solutions with different values of k, $\iota_j(D)$ is the log-likelihood of the dataset D according to model M_j, p_j is the number of parameters in M_j, and R is the number of points in the dataset. X-means chooses the model with the best score.

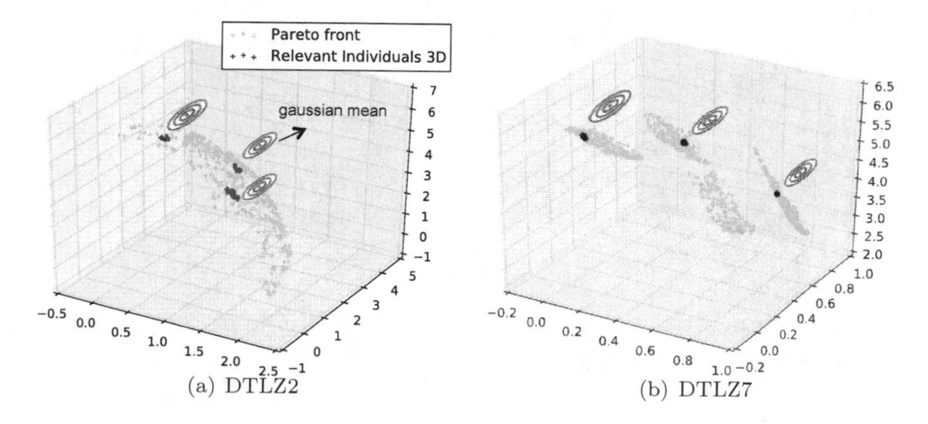

(a) DTLZ2 (b) DTLZ7

FIGURE 3.3: CI-NSGA-II results for DTLZ2 and DTLZ7 problems.

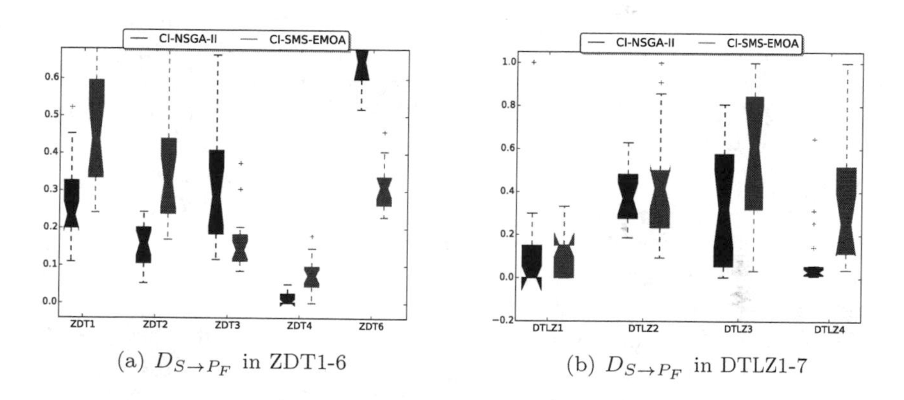

(a) $D_{S \to P_F}$ in ZDT1-6 (b) $D_{S \to P_F}$ in DTLZ1-7

FIGURE 3.4: Distribution of $D_{S \to P_F}$ values for test problems.

In addition to the Gaussian Mixture model, the K-means algorithm was implemented to bring a different clustering technique into the analysis of the algorithms. But the performance of Gaussian Mixture in these cases was consistently better than K-means.

After 30 independent executions per EA on each test problem, Figure 3.5 reports the distribution of the front coverage and dispersion indicators in the form of box plots. For a better visualization, the values were normalized.

Although box plots allow a visual comparison of the results, it is necessary to go beyond reporting the descriptive statistics of the performance indicators. The need for comparing the performance of the algorithms when confronted with the different clustering techniques prompts the use of statistical tools in order to reach a valid judgemnt regarding the quality of the solutions and how different algorithms compare with each other.

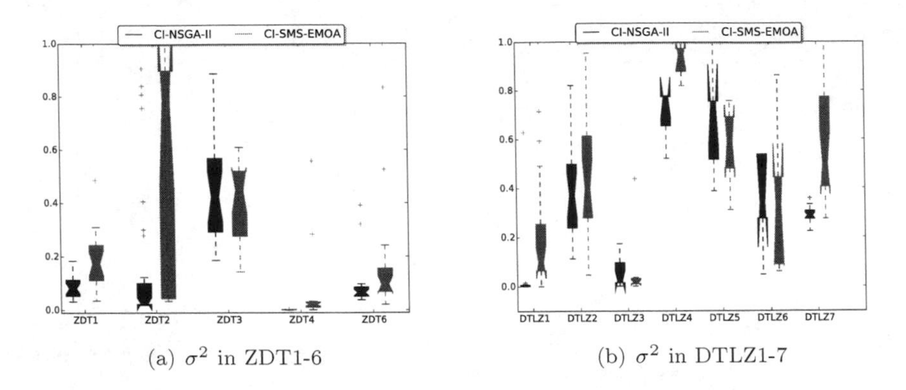

(a) σ^2 in ZDT1-6

(b) σ^2 in DTLZ1-7

FIGURE 3.5: Distribution of $D_{S \to P_F}$ and σ^2 values for test problems.

The Conover–Inman procedure [50] is a non-parametric method particularly suited for this purpose. It can be applied in a pairwise manner to determine if the results of one algorithm were significantly better than those of the other. A significance level, α, of 0.05 was used for all tests. Table 3.1 contains the results of the statistical analysis for all the test problems based on the mean values.

The values of front coverage ($D_{S \to P_F}$) and variance (σ^2) indicators allowed comparison between the new algorithms. The CI-NSGA-II with Gaussian Mixture model consistently outperformed the CI-SMS-EMOA in most cases. It was ranked best in all functions except for ZDT3 and DTLZ1. Considering each of the performance indicators, CI-NSGA-II managed to find a better convergence $D_{S \to P_F}$ in 10 of 12 test problems, losing only in ZDT3 and ZDT6 problems. Regarding the variance indicator (σ^2), the CI-NSGA-II won in all the tests.

In summary, the CI-NSGA-II and its collective reference points proved to be well matched for the range of ZDT and DTLZ test problems.

3.6.2 Resource Distribution Problem

Many companies face problems of resource placement and assignment. The petroleum industry is one of the domain contexts where these problems are present. Those companies must pump oil from resource extraction areas and allocate offshore platforms in such a way that optimizes its operational costs and production of collected resources. This general idea transforms the resource management into a multi-objective problem where one has to operate in an economic way and, at the same time, prioritize the performance or production.

TABLE 3.1: Results of the Conover–Inman statistical hypothesis tests

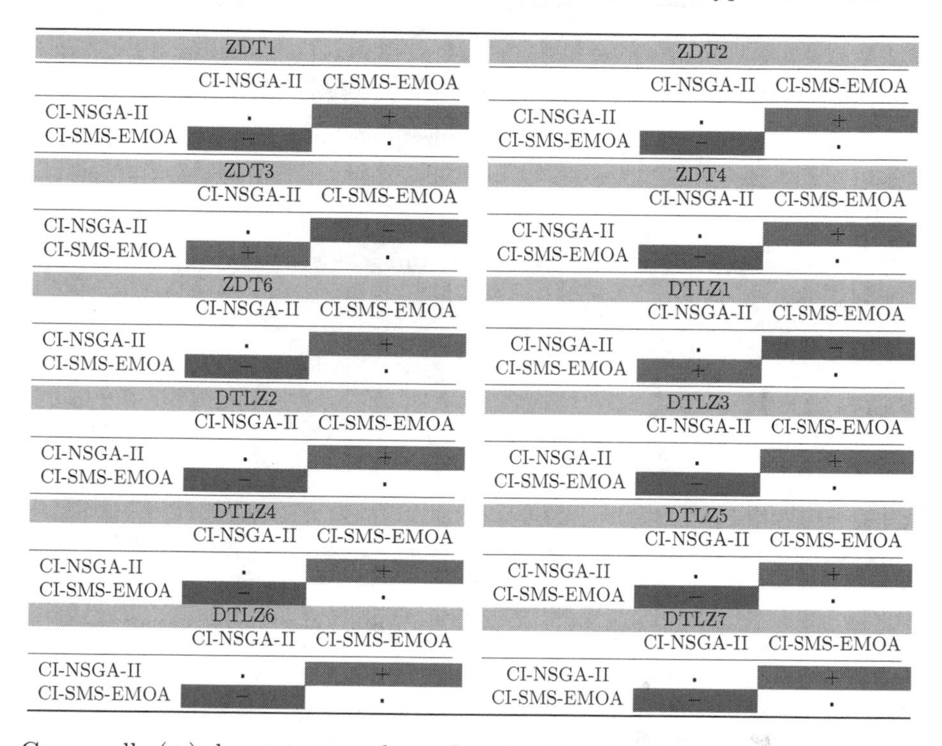

Problem		CI-NSGA-II	CI-SMS-EMOA
ZDT1	CI-NSGA-II	·	+
	CI-SMS-EMOA	−	·
ZDT2	CI-NSGA-II	·	+
	CI-SMS-EMOA	−	·
ZDT3	CI-NSGA-II	·	−
	CI-SMS-EMOA	+	·
ZDT4	CI-NSGA-II	·	+
	CI-SMS-EMOA	−	·
ZDT6	CI-NSGA-II	·	+
	CI-SMS-EMOA	−	·
DTLZ1	CI-NSGA-II	·	−
	CI-SMS-EMOA	+	·
DTLZ2	CI-NSGA-II	·	+
	CI-SMS-EMOA	−	·
DTLZ3	CI-NSGA-II	·	+
	CI-SMS-EMOA	−	·
DTLZ4	CI-NSGA-II	·	+
	CI-SMS-EMOA	−	·
DTLZ5	CI-NSGA-II	·	+
	CI-SMS-EMOA	−	·
DTLZ6	CI-NSGA-II	·	+
	CI-SMS-EMOA	−	·
DTLZ7	CI-NSGA-II	·	+
	CI-SMS-EMOA	−	·

Green cells (+) denote cases where the algorithm in the row was better statistically than the one in the column. Cells marked in red (−) are cases where the method in the column yielded better results statistically when compared to the method in the row.

The problem—to put it in simple terms—has to find a good solution for positioning the processing units according to the resource area. It is formally represented as

$$\min \sum_{i=1}^{N} \sum_{j=1}^{M} \sigma_{ij} d_{ij} + \sum_{j=1}^{M} c_j \mu , \tag{3.7}$$

$$\max \sum_{i=1}^{N} \sum_{j=1}^{M} \sigma_{ij} v_j . \tag{3.8}$$

Let μ be the cost of one processing unit, v the productive capacity of one processing unit linked to one resource area, M a set of available positions to production units, N a set of available positions to resource area, and D a distance matrix $(d_{ef})_{n_x m}$, where $n \in N$ and $m \in M$. The decision variables are the processing unit c_j ($j \in M$) that assumes 1 if it is placed at position j

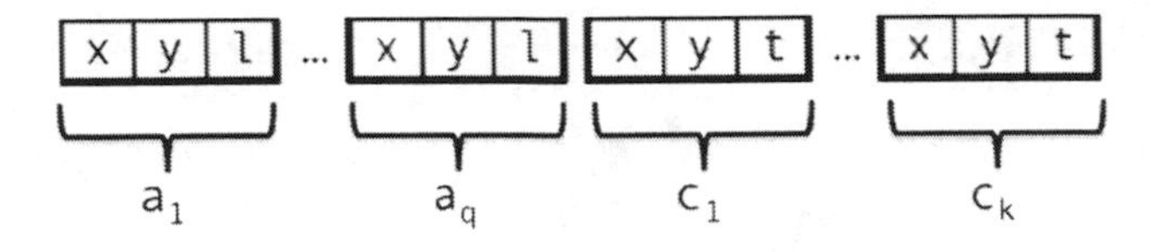

FIGURE 3.6: Chromosome encoding.

or 0 otherwise and σ_{ij} that assumes 1 if there is a link between the resource area at position $i \in N$ and the processing unit at position $j \in M$.

The processing unit is computationally represented as a tuple $c_i =< x, y, t >$, where $c_i \in C = \{c_1, ..., c_k\}$, t is the type of the unit, and x and y are the Cartesian coordinates of the position. The resource area is represented by the tuple $a_i =< x, y, l >$, where $a_i \in A = \{a_1, ..., a_q\}$, and l is a index that links the resource area a_i to the unit c_i. Thus, the chromosome encoding is the aggregation of these tuples regulated by q resource areas and k processing units.

Different constraints from real life and several new interdependencies among the variables might increase the search complexity of this MOP. Progressive articulation of preferences and collective intelligence can implement a dynamism not managed by *a priori* methods and enhance its efficiency. Therefore, the problem described is a candidate for this experiment due to some reasons: a) the objectives and decision variables are meaningful to the group, and the problem is intuitive and allows an interaction with the crowd's cognition; b) incentive engines and gamification can be used to retain the users' interest on the interaction during the optimization; c) the problem can be decomposed in small blocks to be presented to the participants; d) the users' feedback can be parallelized in synchrony with the evolution of individuals in a MOEA.

The experiment was applied in the computer lab of a Brazilian undergraduate institution with more than 30 students' attendance. A gamification feature was implemented to motivate and promote the necessary alignment of goals to all the users in the collectivity. Gamification [78] is the integration of game design elements and game engines in non-game contexts. This is usually intended to increase engagement of players, create gameful and playful user experiences, motivate them, and set clear objectives to guide a cooperative or competitive behavior.

In this context, the resource distribution problem was designed as a game where every player competes among themselves to obtain points and recognition of success. After a certain number of iterations, the CI-NSGA-II algorithm interrupts the evolution process and asks the players for preferred individuals. As votes on the resource distribution scenarios happen, the Gaussian Mixture model calculates the collective reference point to restrict the search to relevant areas in Pareto front. The players who have chosen the individuals near the collective reference point receive a higher score. They compete at ev-

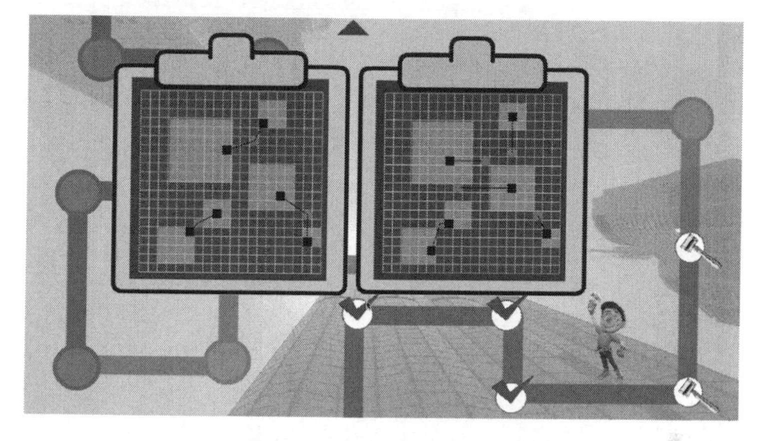

FIGURE 3.7: Gamification features and pairwise comparisons.

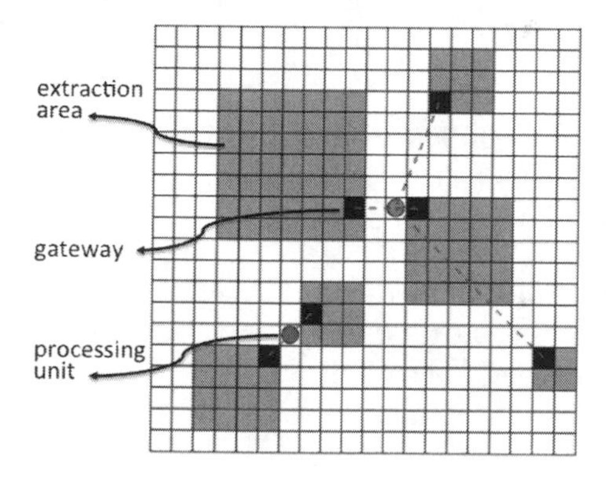

FIGURE 3.8: Evolutionary algorithm solution for six areas.

ery evolution interval for choices around the collective mean. The game needs seven collective interactions to reach optimum points in the front. Figure 3.7 exhibits different phases of the game and the screen for pairwise comparisons of two individuals.

This gamification aims to enhance the results of multi-objective problems through the integration of collective intelligence. At the end of each game, only one scenario progressively created with the support of users' subjectivity and perception is presented to the players. From the group's point of view, it is the best alternative (winner candidate) and overcomes many other optimal points in the front.

Figure 3.8 shows one final solution from a single experiment run. After the collective reference points in a game composed of six resource areas and two processing units, this candidate was progressively created with the support of users' subjectivity and perception.

3.7 Final Remarks

In this work we have discussed group preferences in multi-objective optimization evolutionary algorithms and have introduced a novel approach that brings human subjectivity and cognition into the optimization process. MOEAs can take advantage of decision makers' preferences to guide the search through relevant regions of Pareto-optimal front. The CI-NSGA-II and CI-SMS-EMOA were presented as an interactive approach supported by dynamic group preferences.

The new algorithms have been tested successfully in benchmarking problems and against a real-world case study. Results outlined the benefits of collective reference points to drive the optimization and explore only relevant areas at a trade-off front. The injection of COIN within the original NSGA-II and SMS-EMOA did not make radical changes in their structure. Therefore, the research community should be able to improve and extend the actual achievements.

In the near future, we plan to explore different features of the evolutionary process. We are particularly interested in the usage of COIN as a local search for new individuals during the evolution. Following the Find-Fix-Verify method, it will open the population for users' update to augment its quality. Human brains will unfold creative solutions and possibly explore non-explicit objectives hidden in the problem. After the change, the new individuals designed by a group of people can be directly inserted into the next generation.

Chapter 4

Multi-Objective Particle Swarm Optimization Fuzzy Gain Scheduling Control

Edson B. M. Costa
Federal Institute of Education, Sciences and Technology, Department of Electrical Engineering, Imperatriz - MA, Brazil

Ginalber L. O. Serra
Federal Institute of Education, Sciences and Technology, Department of Electro-electronics, São Luis - MA, Brazil

4.1 Introduction

The pioneer research in gain scheduling control occurred in the early 1950s motivated mainly by the design of autopilots for high-performance aircraft. The main complexities in such projects are the wide range of speeds and altitudes that the aircraft operates, nonlinear dynamics, and time varying characteristics. In general, gain scheduling is a nonlinear control approach that uses a set of linear controllers for various operating conditions of the system. In this approach, one or more measurable variables, called scheduling variables, are used to determine what operating condition the system is currently in and to enable the appropriate linear controller.

Fuzzy control was developed based on the fuzzy set theory proposed by Lotfi A. Zadeh [248] and has been widely applied to various areas including power systems, telecommunications, mechanical/robotic systems, automobile, industrial/chemical processes, aircrafts, motors, medical services, consumer electronics, chaos control, nuclear reactors, and other areas [88]. The theory of fuzzy systems has been widely used in gain scheduling control field applications, giving origin to the field of study known as fuzzy gain scheduling control. Several studies have shown that fuzzy gain scheduling systems have become an important tool to control nonlinear, uncertain, and time varying industrial systems [20, 134, 12, 247, 133, 142].

In this chapter, a fuzzy gain-scheduling controller design based on robust stability criterion via Multi-Objective Particle Swarm Optimization (MOPSO) for Takagi–Sugeno (TS) fuzzy model [230] is proposed. The plant to be controlled is identified from input–output experimental data, by using the fuzzy C-Means clustering algorithm and least-squares estimator for antecedent and consequent parameters estimation, respectively. The MOPSO algorithm is used to tune the fuzzy gain scheduling controller parameters, via Parallel Distributed Compensation (PDC) strategy [239], based on gain and phase margins specifications, according to identified fuzzy model parameters of the plant to be controlled for each operating condition. There are two objectives to be optimized, namely: The first objective is related to obtaining the fuzzy gain scheduling controller parameters to guarantee the gain margin so near as possible of the specified gain margin. The second one is related to obtaining a fuzzy gain scheduling controller parameter to guarantee the phase margin so near as possible of the specified phase margin. These two objectives are conflicting with each other, i.e., when one objective is improved, the other is worsened. Therefore, it is intended to find good "trade-off" solutions subject to a set of constraints that represent the best possible compromises among the gain and phase margins. Experimental results for fuzzy gain scheduling control of a thermal plant with time varying delay are presented to illustrate the efficiency and applicability of the proposed methodology.

This chapter is organized as follows: In Section 4.2 the Takagi–Sugeno fuzzy modeling approach is presented. In Section 4.3 the fuzzy gain scheduling control design approach is presented. In Section 4.4 experimental results for fuzzy gain scheduling control of a thermal plant with time varying delay are shown. Finally, in Section 4.4.3 the conclusions of this work are discussed.

4.2 Takagi–Sugeno Fuzzy Modeling

In this chapter, the fuzzy gain scheduling controller design procedure is based on the representation of a given nonlinear plant in terms of the TS fuzzy model given by:

$$\mathtt{R}^{(i|^{i=1,2,\cdots,L})}: \ \mathtt{IF}\ x_1\ is\ A_1^i\ \mathtt{and}\ \cdots\ \mathtt{and}\ x_\nu\ is\ A_\nu^i\ \mathtt{THEN}$$
$$G_P^i(z) = \frac{b_0^i + b_1^i z^{-1} + \cdots + b_{n_u}^i z^{-n_u}}{1 + a_1^i z^{-1} + a_2^i z^{-2} + \cdots + a_{n_y}^i z^{-n_y}} z^{-\tau_d^i/T} \tag{4.1}$$

where $\mathtt{R}^{(i)}$ denotes the i-th rule, and L is the total number of rules. In the antecedent part, the linguistic variables x_j, $j = 1, 2\cdots, \nu$, belong to a fuzzy set A_j^i with a truth value given by a membership function $\mu_{A_j^i}(x_j) : \mathbb{R} \rightarrow [0,1]$. Each linguistic variable has its own discourse universe $U_{x_1}, \cdots, U_{x_\nu}$, partitioned by fuzzy sets representing its linguistics terms, respectively. The consequent part of the i-th inference rule is composed of n_y-th order discrete-time transfer functions, $G_P^i(z)$, in which τ_d^i is its time pure delay, and $a_{1,2,\cdots,n_y}^i$ and $b_{1,2,\cdots,n_u}^i$ are numerator and denominator parameters, respectively. It can be observed that the TS fuzzy model describes a nonlinear system partitioned into several linear submodels, in terms of transfer functions, forming a convex region (polytope) in the consequent space, as depicted in Figure 4.1.

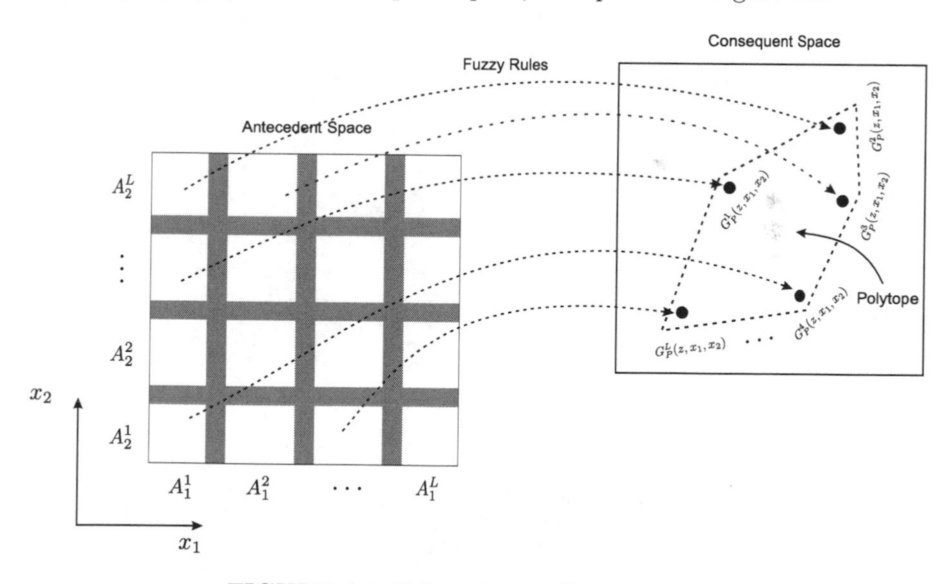

FIGURE 4.1: Polytopic non-linear system.

4.2.1 Antecedent Parameters Estimation

The antecedent parameters of the TS fuzzy model are estimated by fuzzy clustering. The fuzzy clustering algorithms are used to construct fuzzy models from experimental data. Among the most popular algorithms are the following: Fuzzy C-Means (FCM), Gustafson Kessel (GK), and Fuzzy Maximum Likelihood Estimates (FLME) algorithms [25, 126]. The Fuzzy C-Means (FCM) clustering algorithm is used in this chapter, and for more details it can be seen in [24].

4.2.2 Consequent Parameters Estimation

The TS fuzzy model output $\hat{y}$ is computed by taking the weighted average of the individual rules, contributions. Given the normalized degree of fulfillment, γ^i, then the inference formula of the TS fuzzy model in (4.1) can be written as the difference equation

$$\begin{aligned}\hat{y}(k) = & \sum_{i=1}^{l} \gamma^i(k)[-a_1^i y(k-1) - \cdots - a_{n_y}^i y(k-n_y) + \\ & +b_0^i u(k) + \cdots + b_{n_u}^i u(k-n_u)]\end{aligned} \tag{4.2}$$

or can be written as the regression model

$$\hat{y}(k) = \sum_{i=1}^{l} \gamma^i(k)\varphi^T(k-1)\hat{\theta}^i \tag{4.3}$$

where the submodel parameters vector in the i-th rule is given by

$$(\hat{\theta}^i)^T = \begin{bmatrix} a_1^i & a_2^i & \cdots & a_{n_y}^i & b_0^i & b_1^i & \cdots & b_{n_u}^i \end{bmatrix} \tag{4.4}$$

and the regression vector is

$$\begin{aligned}\varphi^T(k-1) = & [-y(k-1) \quad -y(k-2) \quad \cdots \quad -y(k-n_y) \\ & u(k) \quad u(k-1) \quad \cdots \quad u(k-n_u)].\end{aligned} \tag{4.5}$$

In matrix form, considering N input–output pairs of observations $\{u(k), y(k), k = 1, 2, 3, \cdots, N\}$, results

$$\hat{Y} = \Gamma^1\Phi\hat{\theta}^1 + \Gamma^2\Phi\hat{\theta}^2 + \cdots + \Gamma^l\Phi\hat{\theta}^l \tag{4.6}$$

where the matrix Φ is given by

$$\Phi = \begin{bmatrix} \varphi^T(1) \\ \varphi^T(2) \\ \vdots \\ \varphi^T(N) \end{bmatrix} \tag{4.7}$$

the matrix Γ^i is the diagonal weighting matrix of the i-th rule, as follows

$$\Gamma^i = \begin{bmatrix} \gamma^i(1) & 0 & \cdots & 0 \\ 0 & \gamma^i(2) & \cdots & 0 \\ \vdots & \vdots & \ddots & \vdots \\ 0 & 0 & \cdots & \gamma^i(N) \end{bmatrix} \tag{4.8}$$

and $\hat{Y}$ is the output vector of fuzzy model, given by

$$\hat{Y} = [\hat{y}(1) \;\; \hat{y}(2) \;\; \cdots \;\; \hat{y}(N)]^T . \tag{4.9}$$

Considering the output vector of the uncertain dynamic system

$$Y = [y(1) \;\; y(2) \;\; \cdots \;\; y(N)]^T \tag{4.10}$$

then the residual's vector E is defined by

$$E = Y - \hat{Y} \tag{4.11}$$

and the loss function can be written as

$$V(\hat{\theta}, k) = \frac{1}{2} E^T \Gamma^i E = \frac{1}{2} \sum_{i=1}^{k} \left(y(i) - \hat{y}(i)\right)^2 . \tag{4.12}$$

The parameters of the consequent submodels are obtained to minimize the least-squares loss function as follows

$$\hat{\theta}^i = \left[\Phi^T \Gamma^i \Phi\right]^{-1} \Phi^T \Gamma^i Y \tag{4.13}$$

4.3 Fuzzy Gain Scheduling Control

Gain and phase margin specifications are important measures often used as tools for robustness analysis and robust control design to guarantee stability and high performance levels in control systems [6]. Therefore, the gain scheduling fuzzy controller design, in this chapter, is based on the gain and phase margin specifications. The fuzzy gain scheduling controller to be designed based on gain and phase margin specifications presents, without loss of generality, the following $i|^{i=1,2,\ldots,l}$-th rule:

$$\begin{aligned} R^i : \texttt{IF } x_1 \textit{ is } F_1^i \texttt{ AND } \cdots \texttt{ AND } x_q \textit{ is } F_p^i \\ \texttt{THEN } G_c^i(z) = \frac{\alpha^i z^2 + \beta^i z + \gamma^i}{z^2 - z} \end{aligned} \tag{4.14}$$

with

$$\alpha^i = K_P^i + \frac{K_I^i T}{2} + \frac{K_D^i}{T} \tag{4.15}$$

$$\beta^i = \frac{K_I^i T}{2} - K_P^i - \frac{2K_D^i}{T} \tag{4.16}$$

$$\gamma^i = \frac{K_D^i}{T} \tag{4.17}$$

where K_P^i, K_I^i, and K_D^i are the proportional, integral, and derivative gains of local PID controllers, and T is the sample time, respectively [125].

From Equations 4.1 and 4.14, the gain and phase margins of the fuzzy control system are given by:

$$arg\left[\sum_{i=1}^{l}\mu^i G_c^i(z, e^{j\omega_p}) G_P^i(z, e^{j\omega_p})\right] = -\pi \quad (4.18)$$

$$GM = \frac{1}{\left|\sum_{i=1}^{l}\mu^i G_c^i(z, e^{j\omega_p}) G_P^i(z, e^{j\omega_p})\right|} \quad (4.19)$$

$$\left|\sum_{i=1}^{l}\mu^i G_c^i(z, e^{j\omega_g}) G_P^i(z, e^{j\omega_g})\right| = 1 \quad (4.20)$$

$$PM = arg\left[\sum_{i=1}^{l}\mu^i G_c^i(z, e^{j\omega_g}) G_P^i(z, e^{j\omega_g})\right] + \pi \quad (4.21)$$

where the gain margin is given by Equations (4.18) and (4.19), and the phase margin is given by Equations (4.20) and (4.21), respectively. The ω_p is called phase crossover frequency and ω_g is called gain crossover frequency.

4.3.1 MOPSO Based Controller Tuning

The proposed Multi-Objective PSO strategy considered to optimize the parameters α^j, β^j, and γ^j of fuzzy gain scheduling controller in the j-th rule from the gain and phase margins specifications, presents the cost function given by:

$$Cost = \sqrt{\delta_1(GM_c - GM_s)^2 + \delta_2(PM_c - PM_s)^2} \quad (4.22)$$

with

$$\delta_1 + \delta_2 = 1 \quad (4.23)$$

where GM_c and PM_c correspond to gain and phase margin computed; GM_s and PM_s correspond to gain and phase margin specified, respectively.

The particles fly through the search space to find the parameters α^j, β^j, and γ^j that minimize the cost function, given in Equation 4.22. The general flow chart of the multi-objective PSO proposed approach is shown in Figure 4.2.

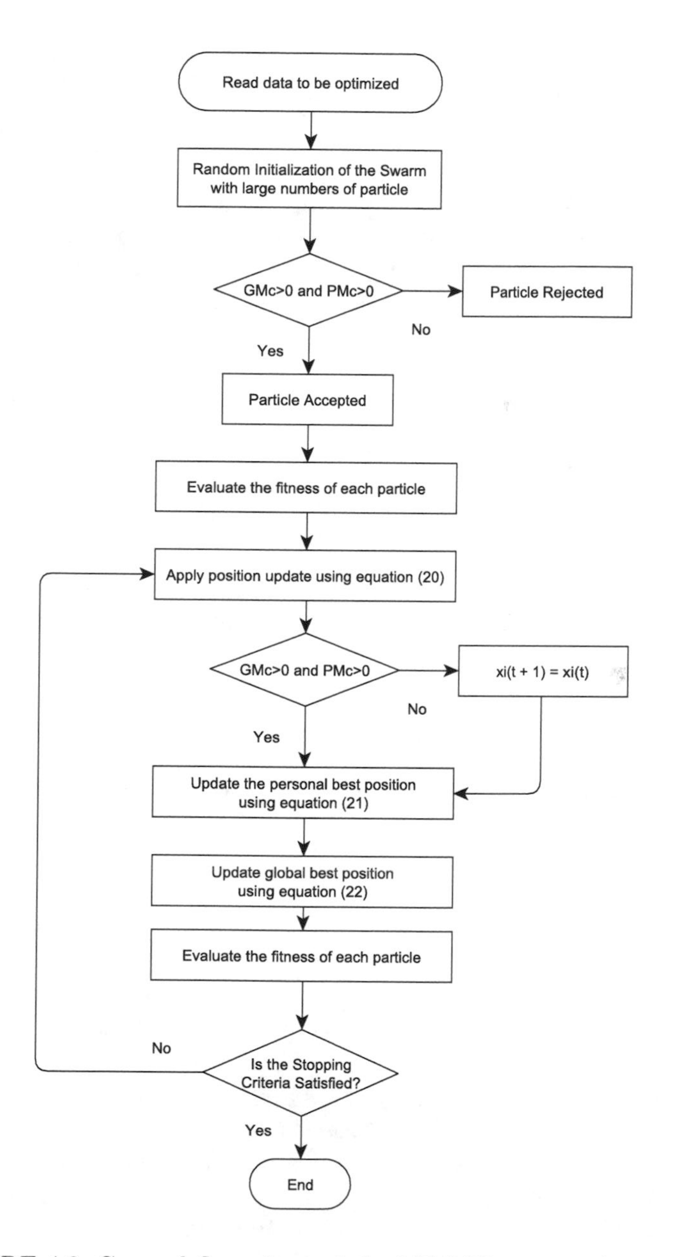

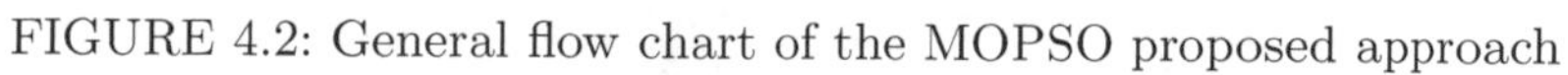

FIGURE 4.2: General flow chart of the MOPSO proposed approach

4.4 Experimental Results

The presented fuzzy gain scheduling control design approach has been tested in real-time experiments applied to a time varying delay nonlinear thermal plant. The data acquisition platform is composed of a virtual instrumentation environment, data acquisition hardware, sensor, and actuator, as shown in Figure 4.3. The thermal plant consists in an adapted monophasic toaster AC 220 volts, with functional temperature from 25 °C to 200 °C. The virtual instrumentation environment (Human Machine Interface) is based in LabVIEW software (LABoratory Virtual Instrument Engineering Workbench) which allows the designer to view, store, and process the acquired data. The data acquisition hardware performs the interface between sensors/actuators and the virtual instrumentation environment, and is composed of the NI cRIO-9073 integrated system, the NI 9219 analog input module, and the NI 9263 analog output module. The temperature sensor is the LM35, and the actuator is based on TCA 785.

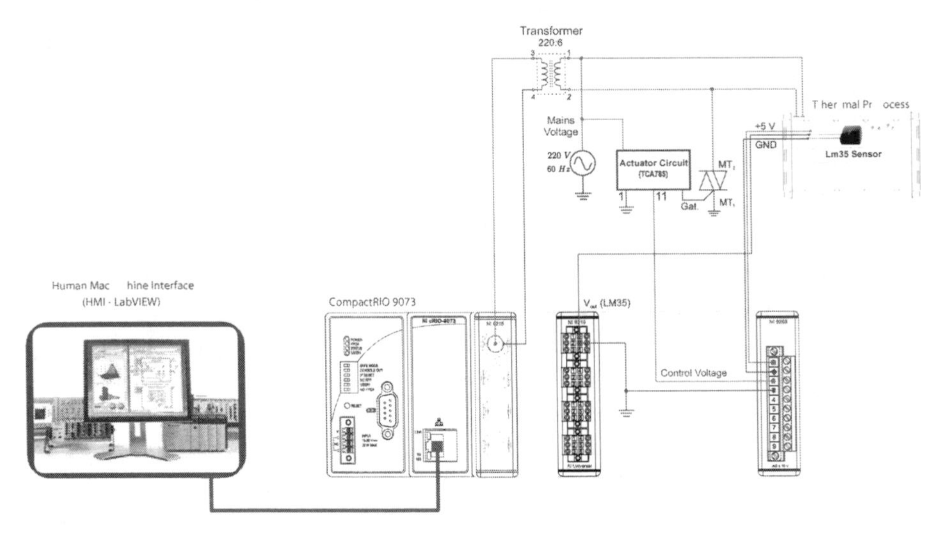

FIGURE 4.3: Data acquisition platform based on virtual instrumentation.

The fuzzy gain scheduling controller design procedure is based on the modeling of two operating conditions of the time varying delay nonlinear thermal plant. The first operating condition corresponds to low temperature values. The second one corresponds to high temperatures. In the sequel, the TS fuzzy modeling procedure and the fuzzy gain scheduling control design of the thermal plant are presented.

4.4.1 TS Fuzzy Modeling of the Thermal Plant

The thermal plant was described by the following fuzzy model structure:

$$R^1 : \texttt{IF } \mathit{Temperature} \texttt{ is } (F^1)^{m_1} \texttt{ THEN}$$
$$G_p^1(z) = k_1 \frac{b_0^1 z + b_1^1}{z^2 - a_1^1 z - a_2^1} z^{-\tau_d^1/T} \tag{4.24}$$
$$R^2 : \texttt{IF } \mathit{Temperature} \texttt{ is } (F^2)^{m_2} \texttt{ THEN}$$
$$G_p^2(z) = k_2 \frac{b_0^2 z + b_1^2}{z^2 - a_1^2 z - a_2^2} z^{-\tau_d^2/T} \tag{4.25}$$

where

$$F^1 = \begin{cases} 1, & y \le p_1 \\ 1 - 2\left(\frac{y-p_1}{p_2-p_1}\right)^2, & p_1 \le y \le \frac{p_1+p_2}{2} \\ 2\left(\frac{y-p_2}{p_2-p_1}\right)^2, & \frac{p_1+p_2}{2} \le y \le p_2 \\ 0, & y \ge p_2 \end{cases} \tag{4.26}$$

and

$$F^2 = \begin{cases} 0, & y \ge p_1 \\ 2\left(\frac{y-p_1}{p_2-p_1}\right)^2, & p_1 \le y \le \frac{p_1+p_2}{2} \\ 1 - 2\left(\frac{y-p_2}{p_2-p_1}\right)^2, & \frac{p_1+p_2}{2} \le y \le p_2 \\ 1, & y \ge p_2 \end{cases} . \tag{4.27}$$

The input signal (RMS voltage, in volts) applied to a thermal plant and its output (Temperature, degree Celsius) obtained for TS fuzzy model identification are shown in Figure 4.4. The antecedent membership functions of a TS fuzzy model (with two rules), given in equations (4.26) and (4.27), were obtained using FCM (Fuzzy C-Means) algorithm and optimized by PSO algorithm: The comparison between the identified and optimized membership functions is shown in Figure 4.5.

The consequent parameters of TS fuzzy model, $\mathbf{G}_p^1(z)$ and $\mathbf{G}_p^2(z)$, given in Equations (4.24) and (4.25), are obtained using the least-square method, as described in Section 4.2.2. The time delay was estimated by computing the cross correlation function between input and output signals of the thermal plant, resulting in a time delay of 130 and 266 samples, corresponding to $\tau_d^1 = 2.210$ seconds and $\tau_d^2 = 4.522$ seconds with sample time of $T = 17\ ms$, for the second order transfer functions in first and second rules, respectively. The parameters of the second order transfer functions in the first and second rules were $b_0^1 = 0.0007$, $b_1^1 = -0.00009$, $a_1^1 = -0.56$, $a_2^1 = -0.44$, and $b_0^2 = 0.03$, $b_1^2 = -0.03$, $a_1^2 = -0.55$, $a_2^2 = -0.45$, respectively.

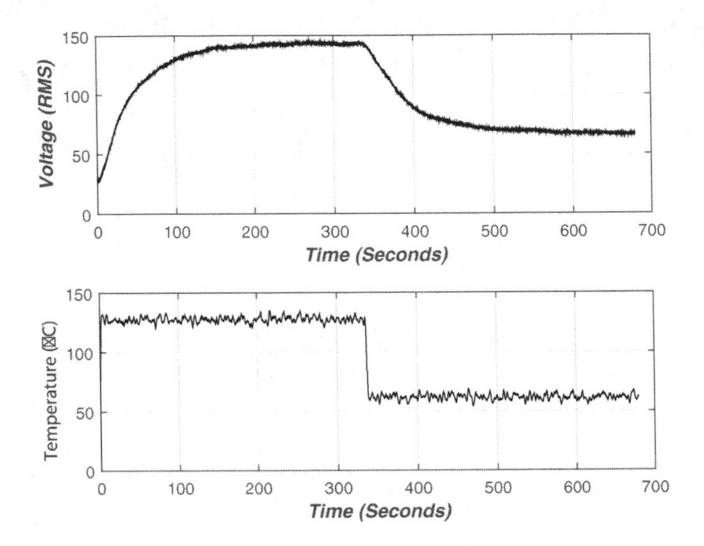

FIGURE 4.4: System identification input (voltage) and output (temperature) signals

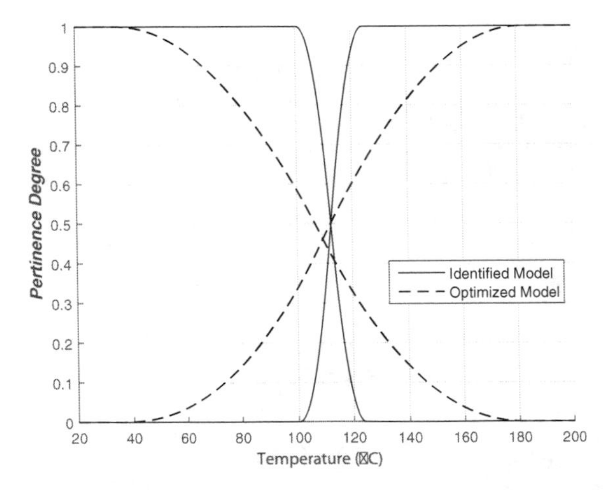

FIGURE 4.5: Estimated membership functions by FCM algorithm (solid line) and the membership functions optimized by PSO (dashed line).

The identified TS fuzzy model was optimized using the PSO algorithm to improve the antecedent membership functions and DC gains of consequent submodels, aiming to minimize the mean squared error (MSE) between the steady-state response of the real process and the optimized model. The objective function to be minimized is given by:

$$e = \frac{1}{Q} \sum_{k=1}^{Q} \left(y_k^i - \hat{y}_k^i \right)^2 \tag{4.28}$$

where y_k and $\hat{y}_k$ are the true and estimated steady-state output, respectively, and Q is the number of points of the characteristic curve of the real process ($Q = 9$). The specified values of PSO parameters were: $s = 50$ (total number of particles in the swarm), $N = 50$ (number of iterations), $c_1 = 1.5$ (acceleration constants cognitive), $c_2 = 1.5$ (acceleration constants cognitive), $w = 0.5$ (inertia weight). The performance of the TS fuzzy model optimized, by PSO, is shown in Figure 4.6. The cost of the best individual in each generation, given by Equation (4.28), is shown in Figure 4.7. The mean squared error of the optimized model is 1.3. And the results obtained for the identified TS fuzzy model and optimized TS fuzzy model are shown in Table 4.1.

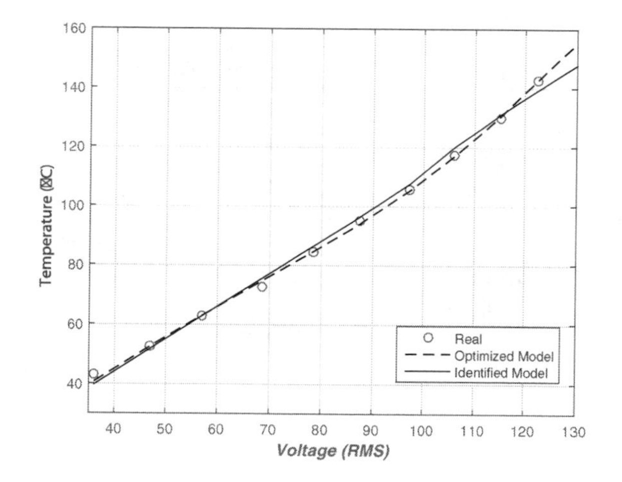

FIGURE 4.6: Static characteristic: thermal plant ("o" line), identified (solid line), and TS fuzzy model optimized by PSO (dashed line).

TABLE 4.1: Parameters and costs for identified and optimized dynamic TS fuzzy models

	Antecedent				Consequent		Cost
Parameters	p_1	p_2	m_1	m_2	k^1	k^2	MSE
Identified model	100.4	124.3	1.0	1.0	1.0	1, 0	22.7953
Optimized model	34.9	181.65	1.0689	1.1698	1.03	1.08	1.4081

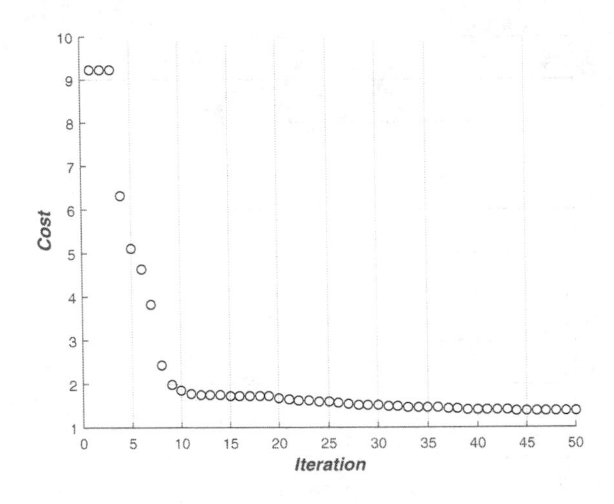

FIGURE 4.7: Cost of the best individual in each generation.

4.4.2 Fuzzy Gain Scheduling Control of the Thermal Plant

From the multi-objective PSO strategy proposed in this chapter, specifying the appropriate gain and phase margins for the fuzzy control system, the parameters of the fuzzy gain scheduling controller resulted in $\alpha^1 = 8.4549$, $\beta^1 = -8.4609$, $\gamma^1 = 0.0090$, $\alpha^2 = 2.3668$, $\beta^2 = -3.1469$, $\gamma^2 = 0.0048$. The gain and phase margin specified were 5 and 70 °, respectively, and the gain and phase margin computed are shown in Figure 4.8 and Figure 4.9, for the submodel 1 and submodel 2, respectively. The TS fuzzy controller obtained is given by Equation 4.29.

$$\begin{aligned}
\mathbf{R}^1 \quad &: \quad \text{IF } \tilde{y}(k-1) \text{ is } \mathbf{F}^1_{k|\tilde{y}(k-1)} \text{ THEN} \\
&\quad \mathbf{G}^1_c(z) = \frac{8.4549z^2 - 8.4609z + 0.0090}{z^2 - z} \\
\mathbf{R}^2 \quad &: \quad \text{IF } \tilde{y}(k-1) \text{ is } \mathbf{F}^2_{k|\tilde{y}(k-1)} \text{ THEN} \\
&\quad \mathbf{G}^2_c(z) = \frac{2.3668z^2 - 3.1469z + 0.0048}{z^2 - z}.
\end{aligned} \tag{4.29}$$

In Figure 4.10, the robust TS fuzzy control of the thermal plant in the presence of external disturbance, where the gain of the plant has changed from 1.22 to 1.77 at time 214 seconds, is shown. The fuzzy gain scheduling controller compensates for the variations in the dynamic behavior of the thermal plant in the presence of an external disturbance.

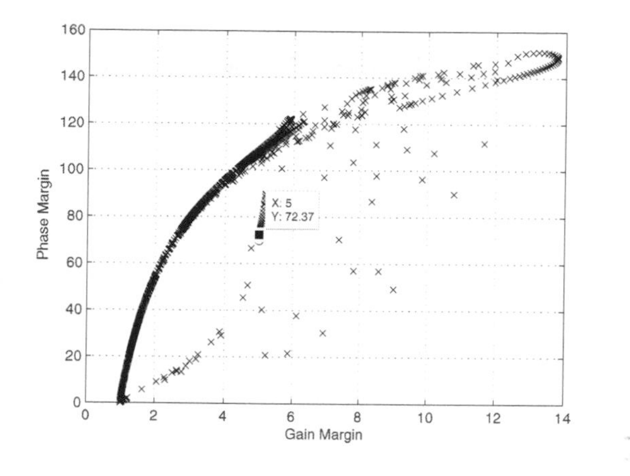

FIGURE 4.8: Gain and phase margin of the first rule.

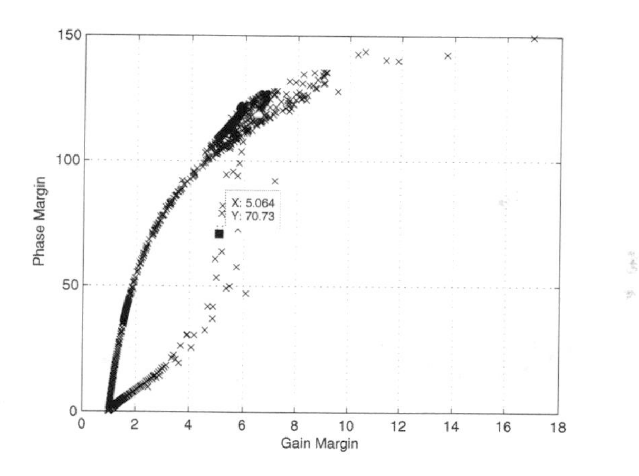

FIGURE 4.9: Gain and phase margin of the second rule.

4.4.3 Final Remarks

In this chapter, fuzzy gain scheduling control approach based on gain and phase margin specifications for nonlinear systems with time varying delay was proposed. The control scheme was efficient to ensure, through the multiobjective PSO approach, that the design criteria are met, as well as the accuracy of the gain and phase margins obtained by the fuzzy control system, and the tracking of the reference trajectory. Experimental results show the high performance of the proposed methodology for fuzzy gain scheduling control of the thermal plant with time varying delay. From the proposed analysis and design, the following future works are of particular interest: applying the pro-

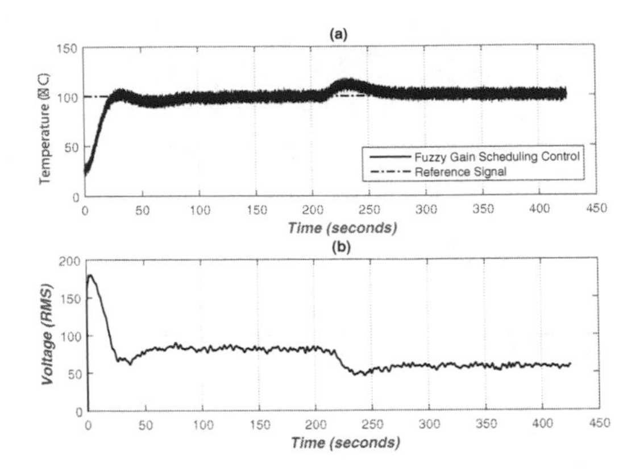

FIGURE 4.10: (a) Temporal response performance of fuzzy gain scheduling controller with external disturbance and (b) Control action obtained with the fuzzy gain scheduling controller.

posed methodology in adaptive control schemes and extending the proposed methodology to MIMO systems.

4.5 Glossary

Gain Scheduling Control: Approach to control of nonlinear systems that uses a family of linear controllers, each of which provides satisfactory control for a different operating point of the system.

Digital PID Control: Control action by controller type PID (with proportional, integral, and derivative gains), in the discrete time domain.

Fuzzy Control: Control based on fuzzy rules. The control action is performed from membership degrees of each controller of rule base.

Robust Control: A control action designed to track a reference trajectory and keep stability, despite uncertainties of the process to be controlled.

Gain Margin: Corresponds to reciprocal of the magnitude of the system, at the phase crossover frequency, i. e., the frequency where the phase angle is $-180°$. It is an important measure of robustness, because it is related to the stability of a closed loop system.

Phase Margin: Is the difference between phase of the system and $-180°$ at the gain crossover frequency, i. e., the frequency where the module of the open loop transfer function is unitary. It is an important measure of robustness, because it is related to the stability of a closed loop system.

Static curve: The static curve of a system corresponds to input and output relation in steady state for various operations points. It can be obtained either from static data (from tests or from historical records), from theory, or from both.

Acknowledgment

The authors would like to thank the CNPQ and FAPEMA for financial support, the master's and PhD Program in Electricity Engineering (PPGEE - UFMA), the Laboratory of Computational Intelligence Applied to Technology (ICAT-IFMA), and the Laboratory of Computational Intelligence and Control (LaCinCo - IFMA) for encouragement and development of this research.

Chapter 5

Multi-Objective Evolutionary Algorithms for Smart Placement of Roadside Units in Vehicular Networks

Renzo Massobrio

University of the Republic, Montevideo, Uruguay

Jamal Toutouh

University of Malaga, Spain

Sergio Nesmachnow

Numerical Center, Computer Science Institute, Engineering Faculty, Uruguay

5.1 Introduction

In the last decades, vehicular traffic has become a major concern in modern cities. This problem is mainly caused by a dramatic increase of the number of vehicles in the roads, which significantly impacts on mobility, road safety, travel times, and environmental issues [130]. Intelligent transportation systems (ITS) have risen from the combination of information and communication technologies (ICT) [107] to solve several problems related to vehicular planning, road mobility, etc. Among different techniques, ITS applies continuous information exchange and computational intelligence [199].

Vehicular ad hoc networks (VANETs) have emerged as one of the most promising solutions to provide network connectivity in vehicular environments to deploy ITS [35, 107]. VANETs are composed of a set of vehicles equipped with *on-board units* (OBU) and *roadside unit* (RSU) elements connected with each other using wireless technologies (see Figure 5.1). Specifically, VANET nodes use direct short range communications (DSRC) to continuously exchange relevant road traffic information with each other. However, the high mobility of the nodes results in frequent changes to the network topology as well as network disconnections, limiting the quality-of-service (QoS) of the system.

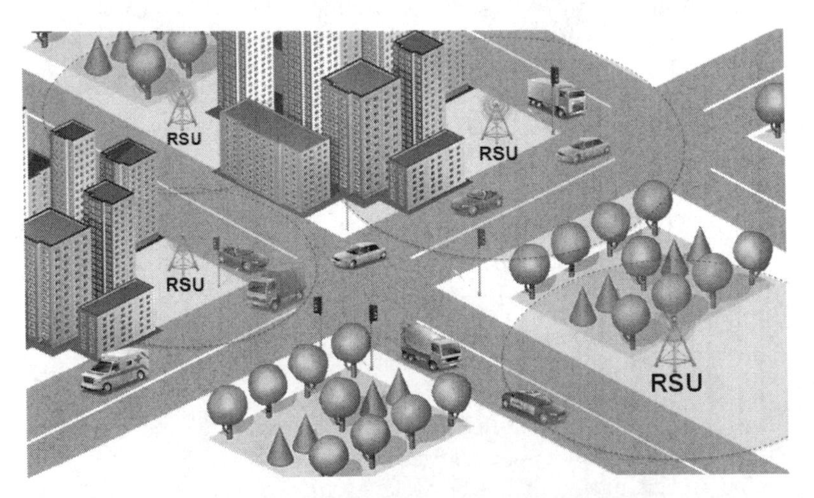

FIGURE 5.1: Basic VANET-based ITS deployed in a city.

There are several VANET services (security, routing, information, entertainment) that use RSUs as information source or receiver. In addition, as

RSU devices have better network capabilities than OBUs, they are used to extend the effective network coverage of the OBUs acting as multi-hop relays. Therefore, installing an efficient RSU infrastructure along the roads (as a VANET backbone) may enhance the QoS of the network.

In this study, we focus on the efficient design of the RSU infrastructure, also known as *RSU Deployment Problem* (RSU-DP), which consists in selecting the best locations and RSU types in order to optimize both the service provided by the fixed infrastructure and the economical deployment costs. This is a tractable problem when dealing with small sized areas, but it results in a hard-to-solve problem for city-scaled instances, as the number of possible solutions becomes very large [209].

In the recent literature there are several studies applying heuristics and metaheuristics to address the efficient design of RSU infrastructures for VANETs [5, 39, 234]. Heuristics and metaheuristics allow traffic engineers to compute competitive solutions (RSU deployments) in reasonable execution times. They apply soft computing search approaches, which are able to tolerate imprecision in data and handle partial solutions in order to build approximate solutions that satisfy the needs of the designers [187]. In this study, we propose applying *multi-objective evolutionary algorithms* (MOEAs) [45, 60] to design an RSU infrastructure to provide VANET services within a city-scaled road network in Málaga, Spain. In order to obtain realistic results, the experimental analysis is performed using real information about road traffic (road map and traffic flow) and hardware (network capabilities and costs).

Three MOEAs are examined and compared to solve the case of study presented in this chapter: a linear aggregation approach and the well-known NSGA-II and SPEA2 algorithms. An important aspect when solving optimization problems using MOEAs is the evaluation of the problem objectives. One of the objectives of the RSU-DP is to minimize the economical cost of the deployment. This objective is evaluated by computing the sum of the cost of each RSU, which depends on its specific type. The other objective in the RSU-DP is to maximize the network service provided by the RSU platform. In this study, a specific QoS model is proposed, considering the number of vehicles, speed, and coverage of street segments in the city, and a Monte Carlo simulation approach is used to compute the corresponding QoS metric.

Thus, the main contributions of the research reported in this chapter are i) introducing a multi-objective formulation of the infrastructure placement problem for VANETs by considering the optimization of the deployment cost and the QoS offered to the users, and ii) an experimental analysis of three MOEAs to solve a specific instance of the problem, built by using real data from the city of Málaga.

The rest of the chapter is organized as follows. Section 5.2 provides an overview about vehicular networks and related technologies. Section 5.3 describes the computational intelligence methods used in this study, introduces the multi-objective optimization concept, and describes the three MOEAs applied to solve the RSU-DP. Section 5.4 presents and formulates the multi-objective RSU-DP addressed in this work and reviews related studies solving different versions of the RSU infrastructure design problem. Section 5.5 details the specific implementation of the evaluated MOEAs to efficiently tackle the RSU-DP. Section 5.6 describes the greedy algorithms used as baseline RSU-DP solvers for comparing the results obtained by the proposed MOEAs, and reports the experimental analysis of the proposed methods on a set of different instances defined using real-world data (Málaga road map and RSU hardware specification). Finally, Section 5.7 states the conclusions and describes the main lines for future work.

5.2 Vehicular Communication Networks

During the last decades, the use of ICT in vehicular environments has prompted the emergence of VANETs. VANETs are networks that connect vehicles equipped with OBUs and RSU with each other by using DSRC technologies [107]. Depending on the type of nodes involved in the communication, two cases are distinguished: i) *vehicle-to-vehicle* (V2V), when two vehicles communicate through ad hoc communications, and ii) *vehicle-to-roadside* (V2R), when the vehicles and roadside elements exchange data with each other.

Based on these types of communications, VANETs can provide a wide variety of powerful applications specifically targeted for vehicles. These applications can improve the safety and efficiency for road users by gathering, processing, and broadcasting real-time traffic information.

In the related literature, VANET applications are categorized into two groups: *safety* and *non-safety* applications. Safety applications use VANETs to exchange relevant information (such as speed, direction, and relative position) to reduce hazardous situations on the road, potentially reducing the number of car crashes. The US Department of Transportation estimates that VANET safety applications have the potential to help drivers avoid or mitigate 80 percent of vehicle crashes [87]. Non-safety applications include a plethora of different services that enhance traffic efficiency (e.g., reduction of travel times or fuel saving), improve passengers' comfort, among others. Nevertheless, these groups are not completely orthogonal. For instance, an application designed to prevent road accidents also improves the efficiency, since it avoids a potential traffic jam the incident may cause.

The vehicular communication system is based on three main components (see Figure 5.2): OBUs, RSUs, and *application units* (AUs). These components are described below:

- OBUs are hardware devices integrated in the vehicles (VANET mobile nodes) to provide them with processing and communication features. Their main functions are i) gathering and processing data collected from the sensors installed in the vehicle and ii) exchanging vehicular information with other VANET nodes (OBUs or RSUs) via IEEE 802.11p DSRC [86]. OBUs may include supplementary network interfaces, e.g., Bluetooth or cellular LTE.

- RSUs are hardware elements installed on the roadside infrastructure elements (e.g., traffic lights, road signals) and on specific dedicated VANET elements located along the roads. They include an IEEE 802.11p network interface to exchange information with other VANET nodes through DSRC. RSUs are also equipped with other network interfaces that provide them with connectivity to other types of networks, such as the Internet. They play three major roles [33]: i) act as an information source or receiver in VANET applications (e.g., warning about the existence of roadworks); ii) extend the effective communication range by forwarding data to other VANET nodes (OBUs or RSUs) through multi-hop communications; and iii) provide Internet connectivity to OBUs.
- AUs may be either external devices connected to a given OBU, such as a smartphone, or a device integrated into the OBU forming a single physical unit. They are connected to the OBU through wired or wireless connection, such as Bluetooth. AUs provide the user interface that allow the user to interact with VANET applications and services.

According to the components involved in the communications, vehicular networks define three communication domains (see Figure 5.2):

- *In-vehicle domain*, which is defined by the OBU, AUs, and sensors connected in a given vehicle. Principally, the OBU uses the network links to gather data from the sensors and share the processed information with the AUs.
- *Ad hoc* or *V2X domain*, which is defined by the OBUs (vehicles) and RSUs forming a mobile ad hoc network (MANET). These nodes exchange information in a fully distributed manner without using any centralized coordination entity through DSRC.
- *Infrastructure domain*, which may include the vehicles' manufacturers, *trusted third parties* (TTP), service providers (SP), and trust authorities (TA). In the infrastructure domain, the RSU serves as a backbone bridge between the infrastructure environment and the ad hoc environment.

In this study we focus on the installation of the required RSU infrastructure along the roads, which is an important problem to be addressed in order to successfully deploy VANET's services.

RSUs are important in VANETs because they operate as road traffic data repositories and information relays for different VANET safety applications. Additionally, they are used to increase the effective coverage and robustness of the vehicular networks and can be also used as a gateway to the Internet,

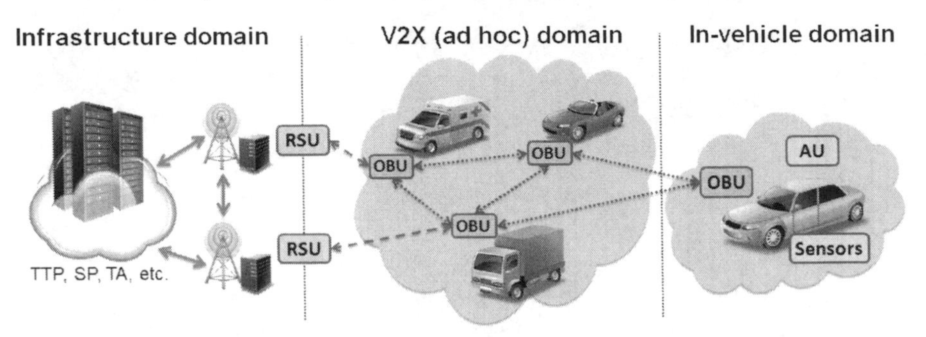

FIGURE 5.2: High-level architecture of VANETs.

allowing data and context information to be collected, stored, and processed in upcoming Cloud infrastructures.

An example of a VANET application that uses RSUs (V2R communication) is the *Intersection Violation Warning (IVW)*, which warns the driver when the possibility of violating a red light is high [107]. A RSU installed in a traffic light controller broadcasts traffic light information including its location, light phase, light timing, and intersection geometry. Then, the OBU installed in the approaching vehicle compares this information with its trajectory and warns the driver if the signal violation is imminent (see Figure 5.3). In addition, the OBU sends to the surrounding VANET nodes a warning message indicating that there is a high probability for a hazardous situation to occur.

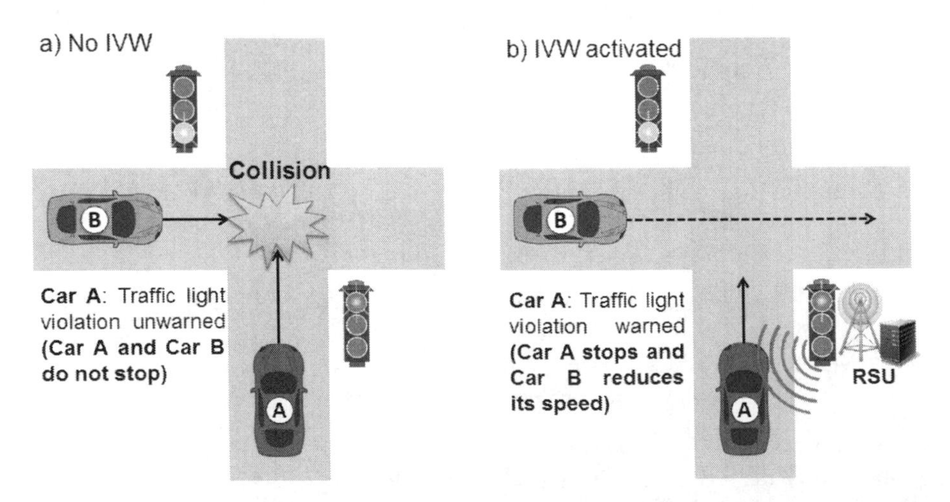

FIGURE 5.3: VANET warning on an intersection violation scenario: (a) without IVW, car A runs the light and collides with car B; (b) when IVW is applied, both cars are alerted and then car A stops and car B can reduce its speed.

5.3 Materials and Methods: Metaheuristics, Evolutionary Computation and Multi-Objective Optimization

This section introduces the main concepts about the methods applied in our research: metaheuristics, evolutionary algorithms, and multi-objective evolutionary algorithms.

5.3.1 Metaheuristics

Metaheuristics are high-level strategies to define algorithmic frameworks that allow designing efficient and accurate techniques to find approximate solutions for search, optimization, and learning problems [98]. They define high-level soft computing methods that can be applied to solve different optimization problems, by instantiating a generic resolution procedure, and needing relatively few modifications to be made in each specific case [187].

In practice, many optimization problems arising in nowadays real-world applications are NP-hard and intrinsically complex. A lot of computing effort is demanded to solve them, due to a number of reasons: they have very large-dimension search spaces, they include hard constraints that make the search space very sparse, they manage very large volumes of data, or they are multimodal or multi-objective problems taking into account hard-to-evaluate optimization functions. This is the case for the problem solved in this chapter: the smart placement of roadside units for vehicular networks.

The smart placement of roadside units for vehicular networks is a special case of the NP-hard problem, because it is a variant of the well-known Radio Network Design problem [174]. Thus, metaheuristics are efficient and accurate methods for solving realistic instances of the problem that often cannot be solved in practice using exact optimization methods (e.g., enumerative search, branch and bound, dynamic/linear/integer programming) that are extremely time-consuming. In this section, we compare the application of multi-objective evolutionary metaheuristics to solve the proposed infrastructure placement problem. The main features of evolutionary algorithms, multi-objective optimization and multi-objective evolutionary algorithms are described next.

5.3.2 Evolutionary Algorithms

Evolutionary algorithms (EAs) are non-deterministic metaheuristic methods that emulate the evolution of species in nature, in order to solve optimization, search, and learning problems [9, 99]. In the past thirty years, EAs have been successfully applied to solve optimization problems underlying many real and complex applications.

The generic schema of an EA is presented in Algorithm 3. An EA is an iterative technique (each iteration simulates a time step, and it is called a *generation*) that works by applying stochastic operators on a set of *individuals* (the population P) in order to improve their *fitness*, a measure related to the objective function that evaluates how good a solution to the problem is. Every individual in the population represents a candidate solution for the problem, according to a specific encoding. The initial population is generated by applying a random method or by using a specific heuristic for the problem (line 2 in Algorithm 3). An evaluation function associates a fitness value to every individual, indicating its suitability to the problem (line 4). The search is guided by a probabilistic selection-of-the-best technique (for both parents and offspring) to tentative solutions of higher quality (line 5). Iteratively, solutions are modified by the probabilistic application of *variation operators* (line 6), most notably including the *recombination* of parts from two individuals or random changes (*mutations*) in their contents, which are applied to building new solutions during the search.

The stopping criterion usually involves a fixed number of generations or execution time, a quality threshold on the best fitness value, or the detection of a stagnation situation. Specific policies are used to select the groups of individuals to recombine (the *selection* method) and to determine which new individuals are inserted in the population in each new generation (the *replacement* criterion). The EA returns the best solution ever found in the iterative process, taking into account the fitness function.

Algorithm 3 Generic schema for an EA

```
1:  t ← 0                                         {generation counter}
2:  initialize(P(0))
3:  while not stopcriterion do
4:      evaluate(P(t))
5:      parents ← selection(P(t))
6:      offspring ← variation operators(parents)
7:      P(t+1) ← replacement(offspring, P(t))
8:      t ← t + 1
9:  end while
10: return best solution ever found
```

One of the most popular variants of EA in the literature is the genetic algorithm (GA), which has been extensively used to solve optimization problems mainly due to its simplicity and versatility.

The classic GA formulation was presented by Goldberg [99]. Based on the generic schema of an EA shown in Algorithm 3 a GA defines selection, recombination, and mutation operators, applying them to the population of potential solutions, which is replaced by the offspring in each generation. In a classic application of a GA, the recombination operator is mainly used to guide the search (by exploiting the characteristics of suitable individuals), while the mutation is used as the operator aimed at providing diversity for exploring different zones of the search space.

5.3.3 Multi-Objective Optimization Problems

Unlike traditional single-objective optimization problems, a Multi-Objective Optimization Problem (MOP) proposes to optimize a group of functions, usually in conflict with each other. As a consequence, there is not a unique solution to the problem, but a *set* of solutions that represent different trade-offs among the optimizing functions' values. A generic formulation of a MOP is:

$$\begin{array}{ll} \text{min/max} & \mathbf{F}(\mathbf{x}) = (f_1(\mathbf{x}), f_2(\mathbf{x}), \ldots, f_k(\mathbf{x})) \\ \text{subject to} & \mathbf{G}(\mathbf{x}) = (g_1(\mathbf{x}), g_2(\mathbf{x}), \ldots, g_j(\mathbf{x})) \geq 0 \\ & \mathbf{x} \in \Omega. \end{array}$$

A MOP solution is a vector of decision variables $\mathbf{x} \in \Omega$ which satisfies the constraints formulated by the functions $\mathbf{G}(\mathbf{x})$, offering adequate trade-off values for the functions $f_i(\mathbf{x})$.

Considering a MOP that proposes to *minimize* a set of objective functions, a solution $\mathbf{w}$ is said to dominate the other solution $\mathbf{v}$ (it is denoted $w \prec v$), if $f_i(\mathbf{v}) \geq f_i(\mathbf{w}) \wedge \exists j / f_j(\mathbf{v}) > f_j(\mathbf{w})$. The set of optimal solutions for a MOP is composed of the non-dominated feasible vectors, named *Pareto optimal set*. It is defined as $P^* = \{\mathbf{x} \in \Omega / \nexists \mathbf{x}' \in \Omega, \mathbf{F}(\mathbf{x}') \preceq \mathbf{F}(\mathbf{x})\}$.

The region of points defined by the optimal Pareto set in the objective function space is known as *Pareto front*, formally defined as $PF^* = \{(f_1(\mathbf{x}), \ldots, f_k(\mathbf{x})), \mathbf{x} \in P^*\}$.

In a MOP there is not a single optimal solution, but a whole set of non-dominated trade-off solutions instead. Because of this, it is very important to apply specific algorithms for MOP solving, which are able to find many solutions in a single execution. The computed solutions should have good quality and be well distributed to sample the Pareto front of the problem. After that, a decision making procedure is needed in order to select which solution(s) will be applied in practice. This decision making is rarely automatized, and it is often performed by a human.

5.3.4 Multi-Objective Evolutionary Algorithms

Rosenberg suggested the capability of EAs for solving MOPs in a pioneering work back in 1967. The first MOEA was presented by Schaffer in 1984. Since 1990, many MOEAs have been proposed by a growing research community that works actively nowadays. These MOEAs have allowed obtaining accurate results when solving difficult real-life optimization problems in many research areas [45, 60].

Unlike many traditional methods for multi-objective optimization, MOEAs find a set with several solutions in a single execution, since they work with a *population* of tentative solutions in each generation. MOEAs must be designed taking into account two goals at the same time: i) to approximate the Pareto front and ii) to maintain diversity instead of converging to a reduced section of the Pareto front, providing a set of different solutions that represent different trade-offs between the problem objectives. A Pareto-based evolutionary search leads to the first goal, while the second one is accomplished by using specific techniques also used in multimodal function optimization (niches, sharing, crowding, etc.). Figure 5.4 graphically presents these two generic goals for a hypothetical problem that proposes to minimize functions $f_1(x)$ and $f_2(x)$.

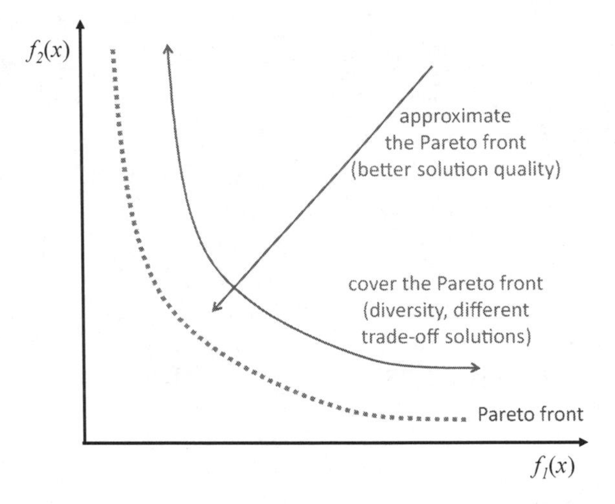

FIGURE 5.4: Goals when designing an MOEA.

In this chapter, we apply three variants of well-known MOEAs:

- *weighted-sum GA using multiple weight combinations* (WSGA).

  WSGA is a traditional evolutionary algorithm that applies a linear aggregation approach to define the fitness function used to evaluate solutions of the problem. This approach is proposed to allow the evolutionary search to focus on different trade-off sections of the Pareto front of the problem, by using several weight combinations.

  In a problem with two objective functions f_1 and f_2, the fitness function in WSGA is defined by the weighted sum of the objective functions of the problem, using a uniformly distributed two-dimensional grid for defining the different weight combinations: $F_{i,j}(x) = w_1^i \times f_1(x) + w_2^j \times f_2(x)$, where $i \in \{1 \ldots M\}, j \in \{1 \ldots N\}$.

  Although the linear aggregation approach can fail to deal with complex multi-objective problems, in practice it is fully applicable to solve a wide range of optimization problems, especially those having convex Pareto fronts [45]. By sampling the search space using different weight combinations, the linear aggregation approach used in WSGA allows sampling the Pareto front of the problem even though the algorithm does not use an explicit fitness assignment schema based on Pareto dominance.

Different variants of this aggregation approach have been successfully applied by the authors in the parallel micro CHC algorithm [188] and also to the classic EA in the parallel Multi-Objective Evolutionary Algorithm with Domain Decomposition [169, 189].

- *Non-dominated Sorting Genetic Algorithm, version II* (NSGA-II) [63]. NSGA-II is a popular state-of-the-art MOEA that has been successfully applied to solve optimization problems in many application areas. NSGA-II includes features to deal with three criticized issues on its predecessor NSGA: i) an improved non-dominated elitist ordering that diminishes the complexity of the dominance check; ii) a crowding technique for diversity preservation; and iii) a new fitness assignment method that considers the crowding distance values.

  A schema of NSGA-II working on a population P (with size N) is presented in Algorithm 4. The fitness calculation is based on Pareto dominance, building *fronts* of solutions, and crowding distance values to evaluate the diversity and the covering of the Pareto front.

Algorithm 4 Schema of the NSGA-II algorithm

```
1:  t ← 0;
2:  offspring ← ∅
3:  initialize(P(0))
4:  while not stopcriterion do
5:      evaluate(P(t))
6:      R ← P(t) ∪ offspring
7:      fronts ← non-dominated sorting(R))
8:      P(t+1) ← ∅; i ← 1
9:      while |P(t + 1)| + |fronts(i)| ≤ N do
10:         crowding distance (fronts(i))
11:         P(t+1) ← P(t+1) ∪ fronts(i)
12:         i ← i+1
13:     end while
14:     sorting by distance (fronts(i))
15:     P(t+1) ← P(t+1) ∪ fronts(i)[1:(N - |P(t+1)|)]
16:     selected ← selection(P(t+1))
17:     offspring ← variation operators(selected)
18:     t ← t + 1
19: end while
20: return computed Pareto front
```

- *Strength Pareto Evolutionary Algorithm, version 2* (SPEA2) [258].

  SPEA2 is a popular state-of-the-art MOEA. It has been successfully applied in many problems in diverse application areas. One of the main distinctive features of SPEA2 is the fitness calculation, which is based on both Pareto dominance and diversity: the algorithm defines the *strength* concept to evaluate how many solutions dominate (and are dominated by) each candidate solution, and a density estimation is also considered for fitness assignment. Elitism is also applied, by using an elite population to store the non-dominated individuals found during the search.

SPEA2 was designed to improve over the main drawbacks of the original SPEA algorithm. The main features of SPEA2 include: i) an improved fitness assignment scheme, taking into account for each individual how many individuals it dominates and it is dominated by; ii) a nearest neighbor density estimation technique which allows a more precise guidance of the search process; and iii) an improved archive truncation method that guarantees the preservation of boundary solutions in the elite population.

A schema of SPEA2 working on a population P (with size N) is presented in Algorithm 5. The elite population is elitePop, having a size of eliteSize. When the elite population is full, a pruning method is applied to remove the most similar individuals to assure that the size of the elite population is always eliteSize.

Algorithm 5 Schema of the SPEA2 algorithm

1: $t \leftarrow 0$;
2: elitePop $\leftarrow \emptyset$
3: **initialize**($P(0)$)
4: **while** not stopcriterion **do**
5:    **evaluate**($P(t)$)
6:    R $\leftarrow P(t) \cup$ elitePop
7:    **for** $s_i \in$ R **do**
8:       $si_{raw} \leftarrow$ computeRawFitness(s_i,R)
9:       $si_{density} \leftarrow$ computeDensity(s_i,R)
10:       $si_{fitness} \leftarrow si_{raw} + si_{density}$
11:    **end for**
12:    elitePop $\leftarrow$ nonDominated(R)
13:    **if** size(elitePoP) $>$ eliteSize **then**
14:       elitePop $\leftarrow$ removeMostSimilar(elitePop)
15:    **end if**
16:    selected $\leftarrow$ **selection**(R)
17:    offspring $\leftarrow$ **variation operators**(selected)
18:    $t \leftarrow t + 1$
19: **end while**
20: **return** computed Pareto front

5.4 RSU Deployment for VANETs

As presented in Section 5.2, including a RSU platform is crucial for the general deployment of VANETs. However, designing such an infrastructure is a challenge in modern cities, because designers must decide about the number, the type, and the location of the RSUs in order to maximize the service provided and network capabilities, while satisfying and/or minimizing the deployment cost constraints. This section presents the RSU-DP formulation proposed and solved in this study. After that, the most relevant studies in the literature about addressing RSU design are reviewed.

5.4.1 The RSU Deployment Problem

The mathematical formulation of the RSU-DP addressed in this study contemplates the following elements:

- The idea is to install a set of RSUs that are defined according to a set of RSU types $T = \{t_1, t_2, \ldots, t_k\}$. In this formulation, the RSU type determines the deployment cost and the communication capabilities (in terms of transmission power and the antenna gain). The type of a RSU is given by the function *type*: $R \rightarrow T$. The deployment cost is computed by the function C: $T \rightarrow \mathbb{R}^+$, where $C(t_h)$ indicates the monetary cost of installing a RSU of type t_h in the deployed RSU infrastructure.

- The RSUs can be located at any place along the considered city streets. Therefore, RSU-DP considers a set of *road segments* $S = \{s_1, \ldots, s_n\}$, which are potential locations for placing a set of RSUs $R = \{R_1, \ldots, R_q\}$. Each segment s_i is defined by a pair of points (p_j, p_k), with $p_j, p_k \in P = \{p_1, p_2, \ldots, p_m\}$. Each point p_j is identified by its geographical coordinates (latitude, longitude). The length of a given segment s_i is given by the function *len*: $S \rightarrow \mathbb{R}^+$. As aforementioned, RSUs can be placed at any location within each segment s_i.

- In addition, an estimation of the road traffic density across each segment s_i in terms of the number of vehicles per time period is given by function NV: $S \rightarrow \mathbb{N}^+$, and the average vehicle speed for each segment is returned by using function *sp*: $S \rightarrow \mathbb{R}^+$.

Solutions to the RSU-DP are defined by a set of RSUs placed over the road segments of the city, represented by a set $sol = \{R_1, R_2, \ldots, R_l\}$ where l (size of sol) is the total number of RSUs in solution sol. In this problem, each RSU is installed in specific coordinates within a segment s_i. However, some road segments might not have any RSU installed on them (l≤n). The segments covered by an RSU are given by the function *cov*: $R \rightarrow S$, and the portion of segment s_i covered by RSU R_j is given by the function *cp*: $R \times S \rightarrow [0, 1]$.

The multi-objective version of the problem proposes to obtain a set of locations and the type of RSU to deploy in each location, with the aim of maximizing the *service time* given by the whole RSU infrastructure, while simultaneously minimizing the *total monetary cost* of deployment. The service time is given by the number of vehicles attended by RSUs and the time they are served (considering the coverage and average speed per each road segment).

Formally, the problem is defined as the simultaneous optimization of two objective functions in conflict with each other: maximize the service time (QoS), given by $f_1(sol)$ (see Equation 5.1), and minimize the cost, computed by $f_2(sol)$ (see Equation 5.2).

$$f_1(sol) = \sum_{R_j}^{R_j \in\, sol} \sum_{s_i \in cov(R_j)} NV(s_i) \times \frac{cp(R_j, s_i) \times len(s_i)}{sp(s_i)}. \tag{5.1}$$

$$f_2(sol) = \sum_{R_j}^{R_j \in\, sol} C(type(R_j)). \tag{5.2}$$

5.4.2 Related Work

The RSU-DP is categorized as a *covering location optimization* problem. It aims at finding optimal positions for RSUs to cover the maximum number of VANET nodes, while taking into account several constraints such as deployment cost, node mobility or application requirements. Several authors have addressed the RSU-DP as a variant of an other well-known covering location problem, the RND problem [174].

The RSU-DP is a hard-to-solve (NP-hard) optimization problem when dealing with (large) instances on city-scaled areas, because the number of possible solutions (i.e., sets of RSU types/locations) becomes very large (virtually impractical) [209]. Therefore, for this kind of real-world instances, traditional exact methods are not able to find accurate solutions in reasonable computation times. For this reason, some authors have analyzed the use of specific heuristics and metaheuristics to deal with RSU-DP.

A common approach that is followed by many researchers is to address the RSU-DP considering only intersections (road junctions) as the best tentative locations to install RSUs. The motivation behind this idea is that the vehicle density at the intersections is usually higher than at other points in the roads [5, 39, 154, 234]. However, following this approach can lead to missing accurate solutions in those cases where installing a RSU in some place between the intersections is the best option. Following a more comprehensive approach, the problem model we consider in this chapter allows to install RSUs in every part of the roads, not just intersections.

A review of the main related works on applying exact methods, heuristic, and metaheuristics to the RSU-DP and related problems is presented next.

5.4.2.1 Exact Methods

Few articles have proposed applying traditional exact approaches to solve the RSU-DP, mainly because the complexity of the problem does not allow to find solutions in reasonable execution times.

In this line of work, Aslam et al. [5] presented a mathematical formulation of the problem, only considering the intersections as possible locations for RSUs. The formulation proposed to minimize the delay time over the routing path for VANET messages, by installing a fixed number of RSUs in a city scenario. The problem model uses information about speed, road traffic density, and the likelihood of hazardous situations for computing the QoS metric (delay time). An exact Binary Integer Programming method was introduced, but the results computed using this traditional approach were outperformed by applying a Balloon Expansion Heuristic algorithm, when solving a problem instance defined in the city of Miami in the United States.

Another proposal of solving the RSU-DP using an exact method was by Liang et al. [151]. However, the proposed resolution method is not able to deal with city-scaled areas; thus, simulation results are presented to evaluate the approach on a university campus map (with a total area of less than 3 km^2) and using randomly simulated traffic data. The problem proposes placing RSUs and selecting their configurations (including antenna types, power level, and wired/wireless network connectivity) to minimize the total cost of deployment and maintenance, subject to user specified constraints on the minimum coverage provided by the RSU infrastructure. A scalability analysis is performed over an area with the size of Cambridge, Massachusetts (about 16 km^2), but no comparison against results computed using other methods is presented.

In a recent proposal, Balouchzahi et al. [13] also addressed the RSU-DP using a binary integer programming formulation oriented to minimize the costs of the RSU deployments. A traditional single-objective approach was applied, and both coverage and QoS considerations are included in the form of optimization constraints. The authors observed that vehicle density has a great impact on the network connectivity and this fact is very important for RSU placement. As in many other problem formulations, the candidate locations for placing RSUs are only the intersections between roads in the studied scenario. The proposed method is implemented in Maple and evaluated over urban scenarios (Erlangen, Zurich, Rome, and New York). Each map is divided in zones (200 m$\times$ 200 m each) to allow the approach a proper scalability behavior to solve large problem instances. However, when using this decomposition approach, the solutions depend on the strategy applied to perform the segmentation. The reported results indicate that the proposed algorithm is able to find better solutions in terms of cost, service advertisement time, and service discovery time, when compared against greedy heuristics from the related literature.

5.4.2.2 Heuristics

One of the first works that explored heuristics to solve a RSU-DP-like problem was the one by Trullols et al. [234]. The authors modeled the RSU-DP as the Maximum Coverage with Time Threshold Problem (MCTTP), to maximize the number of vehicles covered by a given number of RSUs for a given predefined period of time. Three greedy algorithms were proposed, based on performing the search by using information about the road topology and the vehicles. The experimental evaluation was performed over an instance built using real data from Zurich, Switzerland. According to the results, the main finding is that vehicular mobility is the most important factor to take into account when deploying an efficient roadside platform. By taking into account the vehicular mobility information, the proposed greedy heuristics allowed to successfully design a VANET infrastructure capable of informing more than 95% of vehicles in the scenario.

In the article by Xiong et al. [246], the RSU-DP is formulated as a Maximum Coverage Problem (MCP). In the proposed problem model, a given geographical area is studied by applying a division strategy, which considers a grid of non-overlapped zones. Over each zone, traffic data are analyzed over a period of several hours, in order to estimate the number of vehicles entering and leaving each zone for each time unit. This procedure allows generating a mobility directed graph that represents the transition probability for vehicles between two zones in the map. Using the information in the graph, the Minimum Gateway Deployment Problem (MGDP) is formulated to find the candidate zones to deploy RSUs, to be used as gateways to provide Internet access. A specific heuristic algorithm named *MobGDeploy* is introduced to find accurate locations for RSUs/gateways. Although the proposed method tries to cover the maximum number of vehicles by choosing to locate gateways in those zones with high vehicle density, it does not take into account any QoS considerations.

A cost-efficient model for RSU deployment to guarantee certain QoS (delay and success ratio) in a file downloading service was presented by Liu et al. [154]. After applying a theoretical analysis to demonstrate the relationship between the RSU deployment strategy and the file downloading QoS, the chapter formulates the low cost RSU deployment problem as an optimization problem. The road network is modeled as a weighted undirected graph where each edge is the average passing time on the corresponding road. A heuristic ad hoc algorithm based on the depth-first traversal algorithm for edges of a graph is proposed to solve the problem. The simulation results over a generated road network of 4x4 intersections showed that the proposed method can deploy the RSUs along roads with the lowest cost for this size-limited scenario, meanwhile satisfying the user required file-downloading QoS.

A heuristic approach using a Voronoi diagram-based algorithm was applied by Patil and Gokhale [196] to minimize the number of RSUs to deploy in a vehicular network. In the problem model, the main goal is to optimize the network capabilities in terms of packet loss, communications delays, and network coverage, using the lowest possible number of RSUs. The proposed algorithm evaluates information about the speed of vehicles and the traffic density to compute the quality solutions. However, only one type of RSU is considered. The experimental analysis was performed over a medium-size area (about 32 km^2) in the city of Nashville in the United States, using simulated traffic data. The reported results showed that the Voronoi diagram-based placement algorithm was able to compute solutions with less packet delay and less packet loss when compared against two simple placement methods: i) placing RSUs evenly spread across the area and ii) placing RSUs at the busiest intersections. However, the article does not present a comparison against results computed using more powerful placement strategies.

In a more recent article, Ben Brahim et al. [30] solved another variant of the RSU-DP, taking into account real road traffic and mobility data from the city of Doha, in Qatar. Specific pieces of information gathered from the Qatari government were used to define a graph with weighted links to model the problem scenario. In addition to the traditional mobility information (average speed, road traffic density, etc.), Ben Brahim et al. included two other important pieces of information in the problem model to compute the weights of the graphs: the number of hazardous situations and the specific interests on some notable places in the city. A subset of given points in the graph were considered as the potential locations for the RSUs. The problem was addressed by applying two different heuristic approaches: a dynamic algorithm based on 0-1 Knapsack problem solver (*KP_DynAlg*) and the *PageRank* algorithm. The reported results indicate that the PageRank method outperformed the KP_DynAlg algorithm when facing the problem with a given limited number of RSUs. However, in the general case, when the allowed economical budget increased, KP_DynAlg improved the results computed by the PageRank method.

5.4.2.3 Metaheuristics and Evolutionary Computation

Several studies in the literature have analyzed the application of evolutionary metaheuristic algorithms for solving the RSU-DP. The main goal when applying these approaches is to compute accurate and efficient solutions, often improving over those computed using simple heuristic methods, in reasonable execution times.

An early approach applying a GA to solve the RSU-DP was presented by Lochert at al. [156]. The objective of the proposed optimization problem was to maximize the VANET QoS in a highly partitioned network. A VANET simulator was used to evaluate the tentative solutions (i.e., configurations for locating RSUs). A single problem instance was used to evaluate the proposed approach. The scenario was defined by using real data from the city of Brunswick in Germany, including about 500 km of roads and 10,000 vehicles. The authors did not compare their approach against any other methodology for solving the problem. Instead, they assessed the resulting traffic information system and the optimization strategy by means of simulation.

Later, Cavalcante et al. [39] addressed the MCTTP by applying a GA that uses a greedy method to initialize the population. The proposed GA was evaluated over four different instances defined by using real data from cities in Switzerland. The GA results were compared against those computed applying the heuristic methods previously proposed by Trullols et al. [234]. As the main conclusion, the results Cavalcante et al. showed were that the GA solutions guarantee a better vehicle coverage, i.e., up to 11% better than those computed by the greedy approaches.

Another GA to solve the RSU-DP was studied by Cheng et al. [43]. In the proposed GA, the fitness function evaluation took into account the ratio between the road area covered by a given solution and the whole road area. In order to simplify and speed up the computations, the authors used a square grid of $1\,m \times 1\,m$. The GA was evaluated over a problem instance defined in Yukon Territory, Canada, by considering geometry-based coverage information about the roads (without including data related to the mobility of vehicles). The evolutionary approach outperformed the *α-coverage* algorithm, a simple heuristic method that proposes placing the RSUs in the center of the road segments.

Our previous works [168, 170] presented an innovative strategy to address RSU-DP: our papers are the first studies that applied an explicit multi-objective approach in order to consider two different conflicting objectives (the QoS of the VANET and the cost) *to be optimized simultaneously*. Specifically, our approach takes into account two relevant problem objectives: i) maximizing the coverage in terms of the time that vehicles are connected to the RSUs, and ii) minimizing the monetary cost related to the RSU deployment. Our studies consider real information concerning both traffic (speed, traffic density, and road map) from the city of Málaga, in Spain. Real information about the RSU's hardware (costs and network capabilities) is also included to build a realistic city-scaled scenario. Several problem instances were addressed by using a specific MOEA that includes ad hoc operators. The proposed evolutionary approach significantly improved the results computed using ad hoc greedy methods from the literature.

In the study presented in this chapter, we extend our previous conference paper [168] by considering a set of MOEAs that represent a heterogeneous set of different multi-objective solvers, in order to evaluate the capabilities of each evolutionary method to solve the RSU-DP.

5.5 Multi-Objective Evolutionary Algorithms for the RSU-DP

This section describes the features of the proposed MOEAs for the RSU-DP and the greedy heuristics used as a baseline for the comparison of results.

5.5.1 Problem Encoding

In the studied MOEAs, the tentative solutions of RSU-DP are represented as vectors of real numbers. Each vector has a length of n, which is the number of elements in the set of road segments S. Each position on the vector stores the information for the corresponding i segment, i.e., the type and the location of RSU to install (if any). The *RSU type* is represented by the integer part of the real number (0 stands for the absence of RSU in the considered segment, and integers $1 \dots k$ represent types $t_1 \dots t_k$, respectively). The proposed position within the road segment to locate the RSU is given by the fractional part of the real number, mapping the interval $[0, 1)$ to points in the segment $[p_j, p_i)$.

Figure 5.5 presents an example of solution encoding for a given problem instance with four segments and three RSUs installed. In this example, the first position of the vector corresponds to segment s_1, where a RSU of type 2 (integer part of 2.16) is installed at $0.16 \times len(s_1)$ within segment (fractional part of 2.16) $s_1 = (p_1, p_2)$. In the second position, the value 1.50 means that the solution proposes to install a RSU of type 1 at the middle of segment $s_2 = (p_2, p_3)$. Finally, the fourth position of the vector (0.33) shows that no RSU is installed within segment $s_4 = (p_1, p_4)$ (the fractional part of the value encoded is irrelevant in this case).

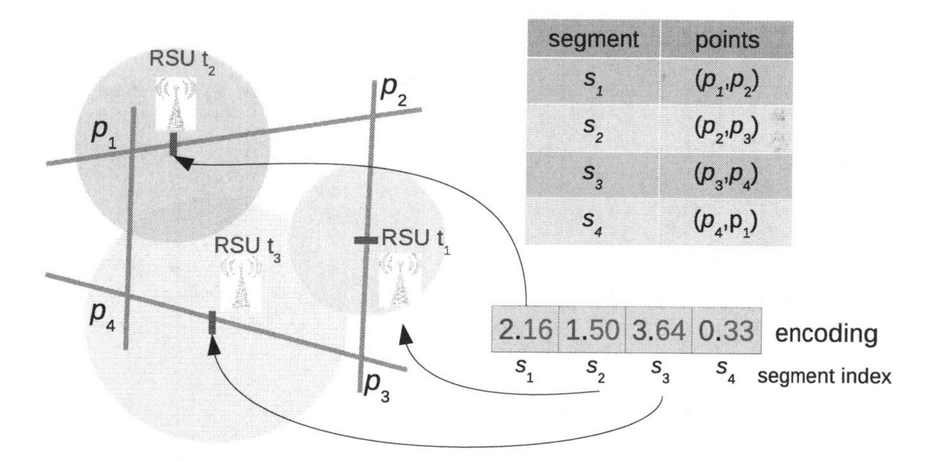

FIGURE 5.5: Proposed encoding for RSU-DP solutions.

5.5.2 Evolutionary Operators

Initialization. In this study, 20% of the solutions of the initial population are seeded by using two randomized greedy heuristics used to compute approximate solutions to the RSU-DP (see a description of the heuristics in Section 5.6.2). This strategy allows the MOEAs to start the evolutionary search from a subspace of good quality solutions. Additionally, the initial population includes the extreme solution of the ideal Pareto front that represents the solution of the minimum monetary deployment cost (i.e., the solution that places no RSU has cost 0). The remaining individuals of this population are initialized by using random real values from the interval $[0, k+r]$ with k the number of different RSU types in T, and $r \in [0, 1)$.

Selection. A tournament selection is applied in the three studied MOEAs. The tournament selects two individuals and the best one survives. Applying tournament selection was originally proposed in NSGA-II and SPEA2 [63, 258] to guarantee the preservation of appropriate features of good solutions during the evolutionary search.

Recombination. The *Intermediate Recombination* (IR) operator is applied with a probability p_C in order to recombine genetic information of two solutions or parents and generate two new solutions or offspring. The IR operator is applied over $\overrightarrow{P1} = \{P1_i\}$ and $\overrightarrow{P2} = \{P2_i\}$, which are combined to create offspring $O1$ and $O2$. These new solutions satisfy:

$$O1_i = \alpha_i P1_i + (1 - \alpha_i) P2_i. \tag{5.3}$$

$$O2_i = \beta_i P2_i + (1 - \beta_i) P1_i. \tag{5.4}$$

In Equations 5.3 and 5.4, α_i and β_i are real numbers randomly chosen from the interval $[-\mu, 1+\mu]$ for a given value of parameter $\mu \in [0, 1]$.

Mutation. The MOEAs analyzed in this study apply an ad hoc mutation operator specifically designed to address the RSU-DP in order to provide diversity during the search process. This operator is applied with probability p_M. When it is performed, the mutation starts by selecting a number of segments to modify (s_i) according to a uniform probability. Then, it applies one of the following three variations over each s_i segment according to a given probability:

- the mutation operator sets the integer part of the selected gene value to 0, thus removing the RSU (if any) from the corresponding segment (applied with probability π_A, see Figure 5.6.a);
- the mutation changes the type of the RSU (if any) to a random type picked uniformly from set T, thus changing the type of the RSU (or adding one if there was none) (applied with probability π_B, see Figure 5.6.b);
- a *Gaussian mutation* on the value for segment s_i is applied with a standard deviation given by parameter σ in order to change the position of the RSU within the segment (applied with probability $1 - \pi_A - \pi_B$, see Figure 5.6.c).

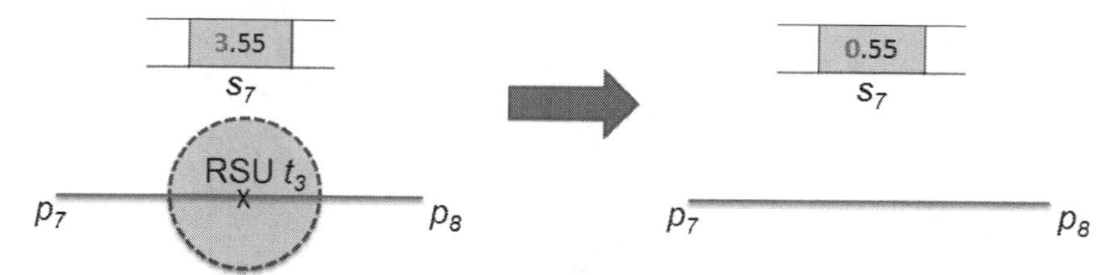

a) Removing a RSU (applied with probability π_A).

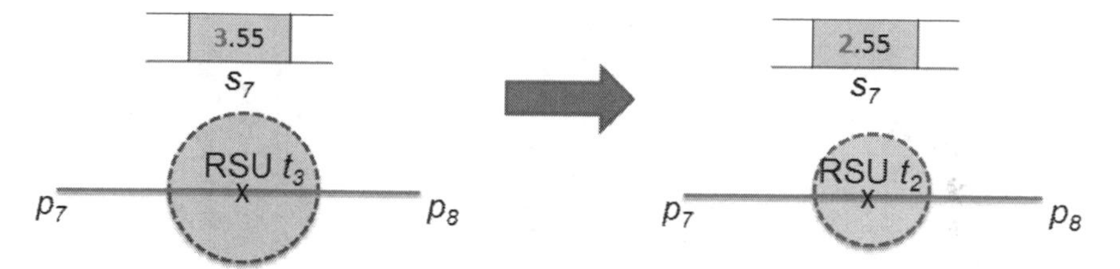

b) Changing the RSU type or adding a RSU (applied with probability π_B).

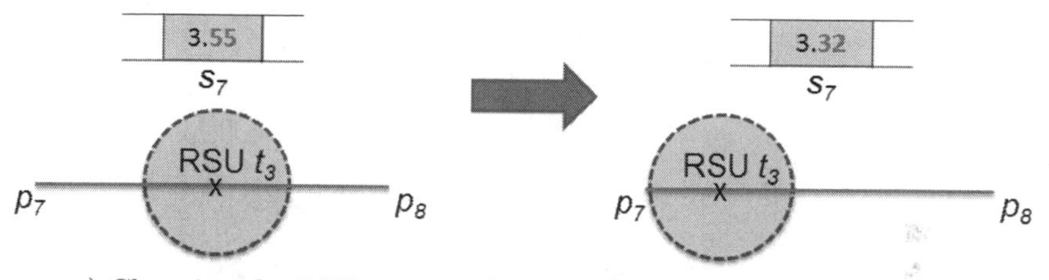

c) Changing the RSU position (applied with probability $1 - \pi_A - \pi_B$).

FIGURE 5.6: Graphical representation of RSU-DP mutation operator.

5.5.3 Evaluation of the Objective Functions

Total cost. The calculation of the total cost is straightforward, by simply adding the cost (according to the corresponding type) of each RSU placed in the scenario.

Quality of Service. For computing the QoS metric (the number of vehicles effectively attended by the RSU infrastructure), we consider the distances and values depicted in Figure 5.7. The RSU placed in segment $s_1 = (p_1, p_2)$ covers the subsegments c_1 (in s_1), c_2 (in s_2) in street A, and c_3 (in s_3), c_4 (in s_4) in street B, according to the coverage defined by the RSU type. The number of effective vehicles attended is given by Equation 5.5.

$$\sum_{i=1}^{i=4} NV(s_i) \times \frac{c_i}{sp(s_i)} \tag{5.5}$$

The QoS calculation requires computing the intersections between the road

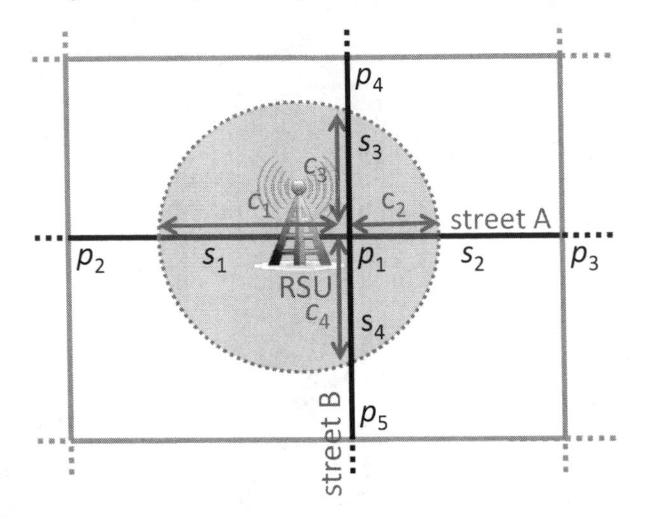

FIGURE 5.7: Calculation of the vehicles attended by a RSU.

segments and the circle representing the coverage of the RSU. We apply a Monte Carlo simulation approach to compute the length of the subsegment c_i, dividing each segment using 10 equally spaced points and considering the coverage radius of the corresponding RSU.

Given that the distances involved in the problem are relatively small, we can estimate them using straight lines in the latitude–longitude space with negligible error, instead of using the Great Circle Distance [244]. Thus, we use the Euclidean distance in the latitude–longitude space as an estimation. This approximation makes the QoS evaluation significantly faster to compute, thus improving the overall performance of the algorithm. Since the distance of a degree of longitude depends on the latitude, it is necessary to adjust for that by multiplying the longitude by the cosine of the latitude for the corresponding city scenario.

5.6 Experimental Analysis

This section describes the experimental evaluation of the proposed MOEAs to solve the RSU-DP.

5.6.1 Problem Instances

When solving problems such as the RSU-DP, which have real-life applications, it is important to perform the experimental analysis using realistic data and scenarios. In this study, the proposed problem instances are defined by using real-world data about the road map, road traffic mobility, and RSU's hardware (network interface/antennae).

The map illustrated in Figure 5.8 shows the area of study in the city of Málaga, which covers 42.557 km^2. Over this area, a set of 128 road segments are defined by 106 geographical points. Some important long roads are represented by more than one unique segment (most notably, avenues like *Avenida de Andalucía*, *Avenida de Valle Inclán*, *Avenida de Velázquez*, and *Paseo Marítimo Pablo Ruiz Picasso*). The road segments have different lengths that vary between 55.5 and 1248.2 m. The average segment length is 483.9 m. All major traffic ways in Malaga, including highways, avenues, and important streets are sampled.

FIGURE 5.8: Map of the area of Málaga and the road segments defined.

The road traffic data are publicly available at the Málaga Council Mobility website [180]. These data were collected by using a set of fixed sensors installed along the roads. These sensors counted the amount of vehicles that circulated between January and June of 2015.

The RSU's hardware is defined by a processing unit equipped with an IEEE 802.11p network interface to perform DSRC communications. Each network interface is connected to an external antenna to extend its communication range and to improve its network capabilities. These antennae are principally characterized by their gain measured in decibels (dB), which indicates the power of the radio signal radiating from the antenna. Therefore, the higher the gain of an antenna, the longer the radio range that can be obtained, and the better the QoS provided by the infrastructure.

In our study, three different types of commercial IEEE 802.11p antennae are considered, according to those available at Cetacea's online shop [40]. Table 5.1 summarizes the main features of these antennae. Thus, the different RSU types considered vary in their communication range and their price.

TABLE 5.1: Antennae used to define the different RSU types

Type	*Commercial model*	*Gain*	*ERR*	*Cost*
t_1	Echo Series Omni Site Antenna	6 dB	243.12 m	121.70 $
t_2	Echo Series Omni Site Antenna	9 dB	338.70 m	139.20 $
t_3	Echo Series Omni Site Antenna	12 dB	503.93 m	227.50 $

In order to evaluate the performance of a given RSU type, we have evaluated its *effective radio range* (ERR). The EER measures the farthest distance at which the RSU may exchange data packets with the vehicles, while guaranteeing a given QoS. The evaluation of the ERR of each RSU type has been carried out by performing realistic VANET simulations, in which the packet delivery ratio (PDR) of a number of V2R communications was evaluated at different distances (from 0 to 650 m). The experiments were performed using the ns-2 simulator [190] to simulate vehicular communications using IEEE 802.11p PHY/MAC standard in an urban scenario defined by a one lane road of 1 km using one RSU and 10 moving cars at 40 km/h. During these simulations, the RSU exchanged data streams at 256 Kbps with the vehicles. In order to obtain realistic results, the non-deterministic fading Nakagami radio propagation model [218] was used to represent the channel characteristics of urban scenarios. The different variants of this VANET scenario were simulated 15 times to compute robust average PDR values. The fourth column of Table 5.1 (*ERR*) summarizes the experimental results by showing the ERR for each RSU type.

5.6.2 Comparison Against Greedy Algorithms

As a baseline to compare the results achieved by the proposed MOEAs, two randomized greedy heuristics are proposed, focused on each of the problem objectives. The greedy algorithms apply intuitive ideas which emulate human-planning strategies for RSU deployment. They are improved versions of the methods defined by Trullols et al. [234] and later used in the comparative study by Cavalcante et al. [39]. The main differences between our heuristics and those presented in [234] are *i*) in our methods, RSUs can be located anywhere within road segments (instead of allowing RSUs to be placed only at road intersections), *ii*) a variable number of RSUs is considered, instead of using a fixed number of RSUs, and *iii*) a set of RSU types with different costs and coverage are considered, instead of a single RSU type. The two greedy heuristics developed are described next:

1. *Greedy QoS.* First, the heuristic sorts the set of segments P according to the QoS they provide in case they are totally covered by a RSU (i.e., the ratio between number of vehicles and average speed). The algorithm iterates over the set of sorted segments, processing them in order (segments with better QoS values are processed first), and selects a random posi-

tion in the segment as a possible location to install a new RSU. Then, two alternatives are considered: *i*) adding to the solution the RSUs that provide the best QoS, in case the QoS can be improved; if two or more RSU types provide the same improvement in QoS, then the cheapest RSU type is selected to be added; or *ii*) not adding a RSU if the QoS cannot be further improved. Segments that are already covered are not taken into account when adding new RSUs.

2. *Greedy cost.* The heuristic starts from the solution computed by greedy QoS, and then it tries to reduce the cost without significantly affecting the quality of service provided by the solution. Different RSU configurations are explored by replacing existing RSUs by cheaper ones or by deleting RSUs, and the option that accounts for the lower QoS degradation is selected. The algorithm stops when all segments are considered or when the QoS of the solution is equal to $\alpha \times Q$, where Q is the best QoS value achieved by the greedy algorithm for QoS and $\alpha \in [0, 1]$. For the experimental analysis, this algorithm was executed using the values $\alpha \in \{0.70, 0.75, 0.80\}$.

The solutions computed by both greedy heuristics tend to group in different regions of the solution space, depending on the parameters used for their execution. Therefore, in order to evaluate the results obtained by the proposed MOEAs, we compare them against the average results for each one of the four identified groups of greedy solutions, which are labeled *gr_1* through *gr_4* in the figures and tables in Section 5.6.4.

5.6.3 MOEAs' Parameter Settings

A parameter analysis was performed in order to find the best values for the parameters in the proposed MOEAs. In the parameter setting experiments, the best results were obtained using the following configuration: *population size* = 72, $p_C = 0.95$, $p_M = 0.01$, $\pi_A = 0.5$, $\pi_B = 0.25$. The value of μ in the *intermediate recombination* operator was set to 0.25. In the Gaussian mutation, the value of parameter σ was set to 0.25. The size of the elite population in SPEA2 was set to 36 individuals. For the weights to use in WSGA, we considered the following sets: $w_{QoS} \in WQ = \{0.0, 0.25, 0.5, 0.75, 1.0\}$, and $w_{cost} \in WC = \{0.0, 0.25, 0.5, 0.75, 1.0\}$. Then, the execution of WSGA was performed using all pairs of weights in the Cartesian product $WQ \times WC$.

5.6.4 Numerical Results

The experimental analysis was performed to assess the problem solving capabilities of the proposed MOEAs for the RSU-DP. We compared the solutions computed by each MOEA against each other and against those computed by the greedy heuristics.

Furthermore, a set of standard metrics for multi-objective optimization [60] were applied in the comparison: *generational distance* (GD), to evaluate the convergence of the computed fronts; *spacing* and *spread*, to evaluate the distributions of solutions; and *relative hypervolume* (RHV) which combines both convergence and dispersion. We also analyzed the Pareto fronts computed by each MOEA in the experimental evaluation over the 20 independent runs that were executed for each of them and for both greedy algorithms.

According to the results obtained in the experimental analysis, NSGA-II showed the best problem solving capabilities among the studied MOEAs. As the reported numerical results demonstrate, NSGA-II significantly outperformed the other two MOEAs as well as the two greedy heuristics, computing accurate Pareto fronts for the problem.

Figure 5.9 shows the global Pareto fronts achieved by the MOEAs as well as the results obtained by the greedy heuristics. Additionally, Figures 5.10 and 5.11 show in more detail the portions of the front where SPEA2 and WSGA achieved its best results, respectively.

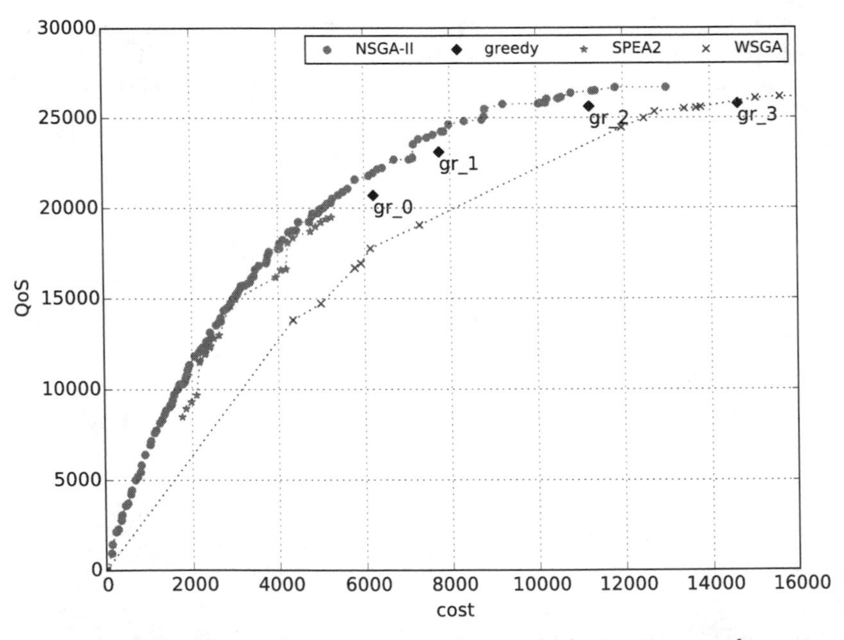

FIGURE 5.9: Global Pareto front and heuristics results.

In Figures 5.9–5.11, it can be observed that NSGA-II is able to find a set of solutions that dominates the solutions found by the other two MOEAs and also the solutions computed by the greedy heuristics. Furthermore, the Pareto front computed by NSGA-II is better both in convergence and diversity than the ones computed by the other two GAs. SPEA2 and WSGA tend to converge

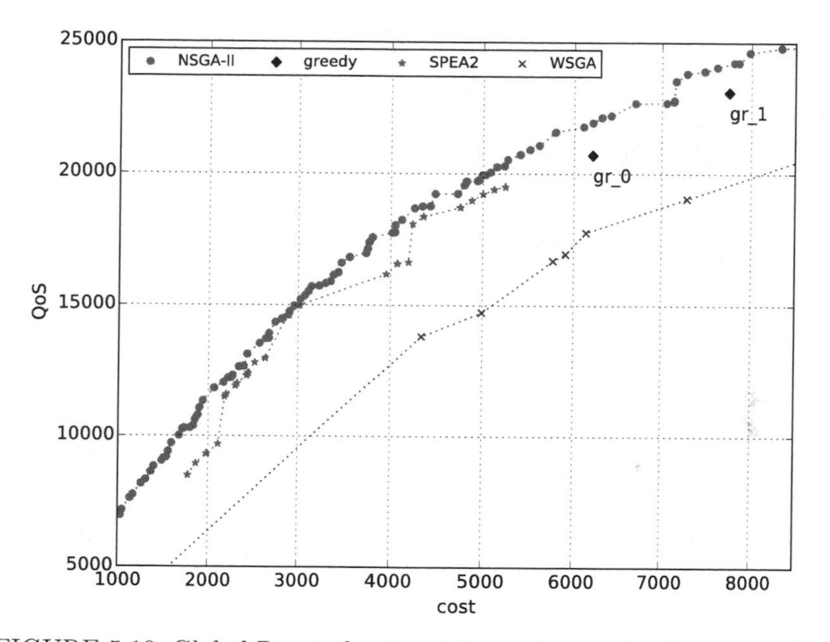

FIGURE 5.10: Global Pareto front and heuristics results (low cost solutions).

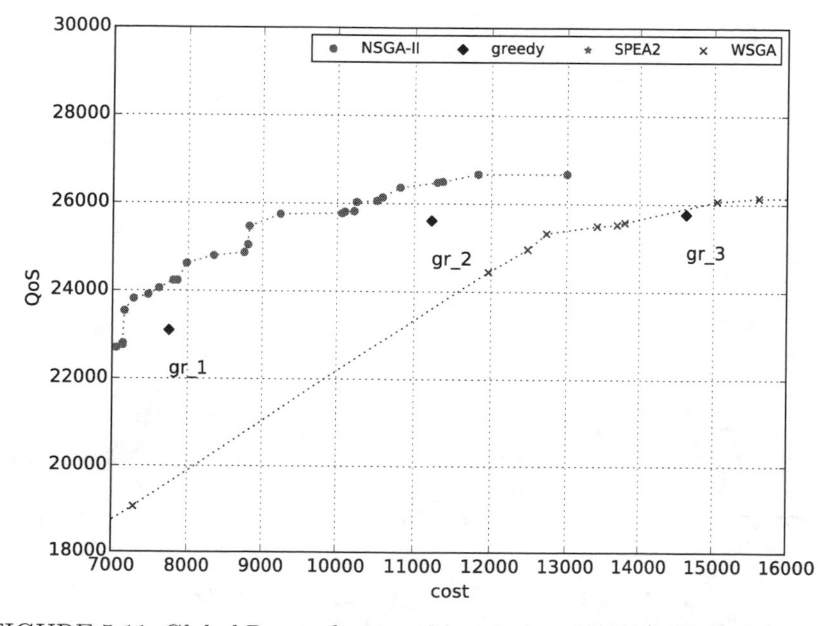

FIGURE 5.11: Global Pareto front and heuristics results (high QoS solutions).

towards different sections of the Pareto front. SPEA2 finds solutions with low costs (and consequently low QoS) while WSGA concentrates on solutions that are more expensive and with greater QoS.

When using MOEAs, the diversity preservation methods play a major role in finding a good set of non-dominated solutions. The $\mu + \lambda$ strategy applied in NSGA-II seems to maintain a good diversity of individuals throughout the generations, while the elite population in SPEA2 fail to do so, concentrating on smaller regions of the solution space than the region sampled by NSGA-II. The worst results were achieved by WSGA, which indicates that a weighted sum approach to the problem fails to adequately explore the solution space and that Pareto dominance-based algorithms are a better choice when dealing with the RSU-DP.

NSGA-II is able to improve the QoS of the greedy heuristics up to **6.0%** while keeping the same cost, and improve up to **36.9%** the cost of the greedy heuristics while keeping the same QoS. Regarding the cost objective, NSGA-II improves over the greedy results 20.3% in average. Improvements on QoS are smaller but still significant: 4.2% in average.

To evaluate the improvements over the greedy algorithms considering the trade-off solutions found for all MOEAs, we take into account the compromise solution (i.e., the solution closest to the ideal vector [60]) computed by each solver. In multi-objective optimization problem, a human decision maker has the final ruling on which of the non-dominated solutions is selected (in our case, which set of RSUs is deployed). The non-dominated set provides different levels of compromise between each objective. The compromise solution offers the best balance between both objective functions; so, it is of particular interest for the decision maker.

Table 5.2 reports the improvements of each MOEA over the greedy solutions (as labeled in Figure 5.9). The compromise solution was computed over the global Pareto fronts achieved by each MOEA, but only considering those solutions with QoS higher than 8000. Solutions with lower QoS (with very low deployment costs, but not providing useful QoS) are not interesting to explore in real-life applications.

TABLE 5.2: Improvements of all MOEAs over the greedy strategies

	Cost improvement				*QoS improvement*			
	gr_1	*gr_2*	*gr_3*	*gr_4*	*gr_1*	*gr_2*	*gr_3*	*gr_4*
NSGA-II	−17.2%	6.0%	35.2%	50.2%	15.1%	3.1%	−7.0%	−7.6%
SPEA2	29.8%	43.7%	61.2%	70.2%	−11.3%	−20.5%	−28.3%	−28.7%
WSGA	−92.6%	−54.4%	−6.6%	18.2%	18.1%	5.8%	−4.6%	−5.1%

The results presented in Table 5.2 further emphasize the fact that SPEA2 focuses on solutions having a low deployment cost for the RSU infrastructure, WSGA focuses on solutions with high QoS, and NSGA-II is able to maintain a good balance between both objectives.

Table 5.3 shows the average, standard deviation, and best results for the studied standard multi-objective optimization metrics achieved by each MOEA. The ideal Pareto front (which is unknown for the problem instance studied) is approximated by combining the non-dominated solutions obtained over all the executions performed of all the MOEA, as it is commonly done in this research area [45].

TABLE 5.3: Comparison of multi-objective optimization metrics for the studied MOEAs

	NSGA-II	SPEA2	WSGA
GD	1.5±0.2 (1.2)	10.2±4.7 (0.0)	20.7±0.8 (18.6)
spacing	208.7±21.5 (169.9)	48.3±61.2 (0.0)	8749.8±844.7 (7077.6)
spread	0.4±0.0 (0.3)	1.0±0.0 (0.9)	0.5±0.1 (0.3)
RHV	$1.0\pm4.6\times10^{-3}$ (1.0)	$0.6\pm6.6\times10^{-2}$ (0.6)	$0.4\pm3.8\times10^{-2}$ (0.5)

The good results for the spacing metric achieved by SPEA2 are misleading. Despite the fact that it outperforms the results achieved by NSGA-II, this is due to the small number of non-dominated solutions computed by SPEA2. The spacing metric measures the relative distance between consecutive solutions in the obtained non-dominated set without taking into account the distance to the extreme solutions [60]. Therefore, if the non-dominated set is comprised of few solutions, it is more likely to achieve good spacing values. On the other hand, the spread metric (which takes into account the full Pareto front for the problem, including in the computation the distance to the extremes points of the Pareto front) clearly shows that NSGA-II outperforms both SPEA2 and WSGA. Furthermore, when looking at the RHV, it is clear that NSGA-II outperforms the other two, both in terms of convergence and diversity of the computed solution fronts.

5.7 Final Remarks

This chapter presents the application of multi-objective evolutionary algorithms to solve the problem of designing an efficient roadside infrastructure to support vehicular networks over realistic urban areas.

The multi-objective formulation proposed in this study considers two conflicting objectives: maximizing the service provided (QoS) by the infrastructure and the monetary deployment cost.

A number of MOEAs were applied to address this optimization problem, including WSGA, NSGA-II, and SPEA2. These algorithms include problem-related operators (encoding and ad hoc mutation) to improve their performance in exploring the search space.

The problem instance used to analyze the proposed methodology was defined on a city-scaled area by using real data (road map, traffic density, and average speeds) from the city of Málaga (Spain). In addition, they take into account three types of real commercial antennae.

As a baseline for our experiments, two state-of-the-art greedy algorithms were developed to address the same problem. Each greedy algorithm solves a single objective: one focuses on optimizing the cost and the other one the QoS. They apply intuitive ideas that emulate planning strategies of a human decision maker in order to construct a solution to the RSU-DP. Among the three MOEAs proposed, NSGA-II was the one that achieved better results. In particular, NSGA-II was able to improve the QoS of the greedy heuristics up to **6.0%** and the cost up to **36.9%**. Additionally, the solutions computed by NSGA-II converged towards an ideal Pareto front of the problem while keeping better diversity than the one achieved by both SPEA2 and WSGA.

The main future research areas are related to extending the problem formulation in order to include additional relevant road traffic information, such as locations of car accidents and traffic jams or points of interest in the city (schools, industrial areas, etc.). In addition, we plan to take into account different types of VANET applications in order to model more realistic QoS evaluation. Finally, we are working on building a larger set of real RSU-DP scenarios based on real information and areas from different cities, e.g., Montevideo in Uruguay and Cardiff in Wales, United Kingdom.

Chapter 6

Solving Multi-Objective Problems with MOEA/D and Quasi-Simplex Local Search

Lucas Prestes
Department of Computer Science, Universidade Estadual do Centro-Oeste - UNICENTRO, Guarapuava, Paraná, Brazil

Carolina Almeida
Department of Computer Science, Universidade Estadual do Centro-Oeste - UNICENTRO, Guarapuava, Paraná, Brazil

Richard Gonçalves
Department of Computer Science, Universidade Estadual do Centro-Oeste - UNICENTRO, Guarapuava, Paraná, Brazil

---

6.1 Introduction

Several practical problems can be formulated as optimization problems [228, 46], particularly in the Operational Research area [131, 202, 251]. Some of these problems have multiple conflicting objectives that need to be optimized at the same time, i.e., are Multi-Objective Problems (MOPs). The solution to a MOP is usually composed of a set of solutions that corresponds to different trade-offs between the objectives. In the absence of a priori information, the Pareto dominance concept is commonly used to determine the solutions that belong to this set, which is then called Pareto set.

Some MOPs cannot be solved by exact methods efficiently and, therefore, approximation algorithms, such as Multi-Objective Evolutionary Algorithms (MOEAs), are used to solve these problems. Another feature of MOEAs is their capability to find multiple solutions (a Pareto set approximation) in a single run. Recently, decomposition based MOEAs, such as MOEA/D (Multi-Objective Evolutionary Algorithm based on Decomposition), have become a popular choice for solving MOPs [256, 150]. MOEA/D decomposes a MOP into multiple single objective subproblems that are simultaneously optimized by the collaboration between the solutions for the different subproblems [254]. Although the original version used Genetic Algorithm operators, the most common version of MOEA/D uses Differential Evolution operators to solve the single objective subproblems.

The efficiency of MOEA/D is demonstrated in one of its variants, which won the CEC 2009 Multi-Objective Competition [256]. Despite the good results, there is still opportunity for further improvements of the MOEA/D framework and its variants. So, the main aim of this chapter is to improve the performance of MOEA/D by introducing a local search step into its framework, i.e., by generating better solutions during the evolutionary process. Two local searches are investigated. Both are Quasi-Simplex (Nelder–Mead) variations [79]. The proposed method is called MOEA/D$_{QS}$.

An empirical analysis of the influence of the components and parameters of the local searches in the MOEA/D performance is done and the best version is compared with the original MOEA/D and NSGA-II algorithms, which are two of the most cited and influential MOEAs in the literature. Four benchmarks commonly employed in the Multi-Objective Optimization literature (CEC 2009, WFG, DTLZ and ZDT) are used to evaluate the algorithms. The quality of the Pareto approximations is measured in terms of three distinct quality indicators: hypervolume, addictive unary-ϵ, and IGD.

The organization of the chapter is as follows. Section 6.2 presents the main concepts from Multi-Objective Optimization and Section 6.3 describe MOEA/D. Differential Evolution is presented in Section 6.4. The

Quasi-Simplex local search is described in Section 6.5. Section 6.6 presents the proposed combination of MOEA/D and local search (MOEA/D_{QS}). The results obtained and an analysis of the components and parameters of MOEA/D_{QS} are presented in Section 6.7. Finally, Section 6.8 ends the chapter with the conclusions and future works.

6.2 Multi-objective Optimization Problems

A Multi-Objective Optimization Problem (MOP) has M $(M > 1)$ objective functions to be minimized or mixmized and is formally defined accordingly in Equation 6.1 [65].

$$\begin{array}{ll} \text{Minimize/Maximize } f_m(\mathbf{x}) & m = 1, 2, ..., M; \\ \text{subject to } \; r_g(\mathbf{x}) \geq 0, & g = 1, 2, ..., G; \\ \qquad r_h(\mathbf{x}) = 0, & h = 1, 2..., H; \\ \qquad x_d^{(L_w)} \leq x_d \leq x_d^{(U_p)} & d = 1, 2, ..., n. \end{array} \tag{6.1}$$

A solution minimizes (or maximizes) the components of the objective vector $\mathbf{f}(\mathbf{x})$ where $\mathbf{x}$ is a N-dimensional decision variable vector $\mathbf{x} = (x_1, ..., x_n) \in \Omega$. Lower $(x_d^{(L_w)})$ and upper $(x_d^{(U_p)})$ bounds can be established to each decision variable x_d. These limits constitute the decision variable space (D). The problem has G inequality and H equality constraints, represented by the inequality and equality functions, $r_g(\mathbf{x})$ and $r_h(\mathbf{x})$, respectively. A MOP can be constrained or unconstrained. For unconstrained problems, as those considered in Section 6.7, only the decision variable space D is defined.

Usually, as long as the multiple objectives are conflicting, there is not a single solution that is optimal with respect to all objectives, so the solution of a MOP is a set of optimal solutions. In the absence of a priori preference information, the Pareto optimality concept is used to define the optimal solution set, which is called Pareto set. A solution is Pareto optimal if it is not dominated by any other feasible solution. A solution x dominates a solution y if $\forall i \in \{1, 2, 3, \ldots, M\} : f_i(x) \leq f_i(y) \wedge \exists j \in \{1, 2, 3, \ldots, M\} : f_j(x) < f_j(y)$.

The approaches used to solve MOPs can be classified into four different categories [27]:

- **Pareto-based:** This approach uses the Pareto dominance to guide the convergence and a diversity measure to guarantee good coverage of the Pareto front. NSGA-II [64] and SPEA2 [259] are the most famous and efficient representatives of this category.

- **Indicator-based:** The algorithms from this category use quality indicators (such as the hypervolume and unary-ϵ) to guide the search. IBEA [267], HypE [11], and SMS-EMOA [23] represent this category.

- **Decomposition-based:** This approach transforms a MOP in multiple single objective problems. It is based on the concept that solving a single objective problem is easier than solving a MOP. MOEA/D is the best known decomposition-based algorithm.

- **Component-wise Design:** Algorithms from this category use automatic configuration tools to select algorithmic components from Pareto-based and Indicator-based algorithms. They also find good values for the parameters associated with these components. AutoMOEA [27][26] is currently the only representative of this category.

In this chapter, a decomposition-based approach is used, particularly if our algorithm is a variant of the MOEA/D algorithm with a differential evolution operator and quasi-simplex local search.

6.3 Multi-Objective Evolutionary Algorithm Based on Decomposition

MOEA/D is based on conventional aggregation approaches [83] as it decomposes a MOP into a number of single objective optimization subproblems. The objective of each subproblem is a linear (or nonlinear) weighted aggregation of all individual objectives in the MOP. Neighborhood relations among these subproblems depend on distances among their aggregation weight vectors. MOEA/D, generally, uses a set of M evenly spread weight vectors, where M is the number of subproblems. Each subproblem is simultaneously optimized using mainly information from its neighboring subproblems.

There are various versions of MOEA/D, including one that won the CEC 2009 MOEA contest: MOEA/D-DRA (MOEA/D with Dynamical Resource Allocation) [256]. In order to allocate the computational resources to the most appropriate subproblems, MOEA/D-DRA utilizes a tournament selection based on the utility value of each subproblem (π^i). The utility of each subproblem is calculated according to Equation 6.2.

$$\pi^i = \begin{cases} 1, & \text{if } \Delta^i > 0.001 \\ (0.95 + 0.05 * \Delta^i/0.001) * \pi^i, & \text{otherwise} \end{cases} \tag{6.2}$$

where Δ^i is the relative decrease of the objective function value of subproblem i. The subproblems with greater Δ^i values have better chances of being selected.

MOEA/D and its variants can use any decomposition approach for defining their subproblems. The most commonly used are the Weighted Sum (Equation 6.3), Tchebycheff approach (Equation 6.4), and PBI (Penalty-based Boundary Intersection, Equation 6.5).

$$\text{Max } g^{ws}(\mathbf{x} \mid \boldsymbol{\lambda}, \mathbf{z}^*) = \sum_{i=1}^{M} \lambda_i f_i \quad (6.3)$$
$$\text{subject to } \mathbf{x} \in \Omega$$

where g^{ws} is the Weighted Sum function, $\mathbf{f}(\mathbf{x}) = (f_1(\mathbf{x}), ..., f_M(\mathbf{x}))$ is the multi-objective function to be minimized, and $\boldsymbol{\lambda} = (\lambda_1, ..., \lambda_M)$ is the weight vector associated with subproblem i.

$$\text{Min } g^{te}(\mathbf{x} \mid \boldsymbol{\lambda}, \mathbf{z}^*) = \max_{1 \leq j \leq M} \{\lambda_j \mid f_j(\mathbf{x}) - z_j^* \mid\} \quad (6.4)$$
$$\text{subject to } \mathbf{x} \in \Omega$$

where g^{te} is the Tchebycheff function, $\mathbf{z}^*$ is the empirical ideal point, and the other variables are as previously defined.

$$\text{Min } g^{pbi}(\mathbf{x} \mid \boldsymbol{\lambda}, \mathbf{z}^*) = d_1 + \theta d_2 \quad (6.5)$$
$$\text{subject to } \mathbf{x} \in \Omega$$

where g^{pbi} is PBI function, θ determines intensity of the penalty associated with the distance to the weight vector, and d_1 and d_2 are defined as in Equation 6.6.

$$\begin{aligned} d_1 &= \frac{\|(z^* - F(x))^T \lambda\|}{\|\lambda\|}, \\ d_2 &= \|F(x) - (z^* - d_1 \lambda)\|. \end{aligned} \quad (6.6)$$

The Weighted Sum function is not capable of finding every Pareto optimal solution if the Pareto front is nonconcave [176]. The Tchebycheff function is not smooth while the PBI function needs an additional parameter. Furthermore, one problem with using the empirical ideal point (Tchebycheff and PBI) is the possible concentration of solutions in a specific region of the Pareto front [216].

This work uses the Tchebycheff approach because some benchmark functions have nonconcave Pareto fronts and we didn't want to introduce more parameters to the proposed algorithm.

According to [217] the MOEA/D can be investigated in one of four major research directions. The works associated with the first direction focus on methods to generate new solutions in the MOEA/D framework. An example of work in this direction is [149], where the genetic operators used in the original MOEA/D are substituted by DE operators. The second direction is associated with works that investigate new ways to decompose the MOP into single objective problems and plans to adapt the ways in order to obtain a better

Pareto front approximation. MOEA/D-M2M [153] belongs to this category, as it divides the solutions into regions (each region is a different subproblem). The parent selection is the focus of the third direction and MOEA/D-DRA is the best known example of this category. The last direction is focused on finding better scalarizing functions. For example, the inverted PBI is investigated in [217].

This work focuses on generating better solutions by combining the DE operators and a Quasi-Simplex local search, therefore, according to the classification described in the previous paragraph, the current work can be classified into the first research direction.

6.4 Differential Evolution

Differential Evolution (DE) is a stochastic, population-based search strategy developed by Storn and Price [226]. DE has obtained very good results in nearly all the CEC competitions on single-objective, constrained, dynamic, large-scale, multi-objective, and multi-modal optimization problems [58]. It has the self-referential mutation property, i.e., it scales the mutation intensity according to the dispersion observed in each variable (variables with low dispersion result in small mutation values while variables with high dispersion present high mutation values)[58]. This property improves the search ability of the algorithm.

DE is a simple algorithm that has three control parameters: NP, F, and CR. NP is the population size. The scaling factor parameter (F) is used to scale the difference between vectors, which is further added to a target vector $\hat{\mathbf{x}}$. This process is called *mutation* in DE. The vector resulting from mutation, named *trial vector*, is combined with a parent vector $\mathbf{x}^p$ in the crossover operation, according to the Crossover Rate parameter (CR). Finally, the offspring is compared with its parent vector to decide (based on their fitness) who will "survive" to the next generation.

There are some variations to the basic DE, they differ especially in the way the target vector is selected (x), the number of difference vectors used (y), and the way that the crossover point is determined (z). The notation adopted to characterize the variations is DE/x/y/z. According to [58], the five most cited DE variants are:

- DE/rand/1: $V_{i,Gen} = X_{R1,Gen} + F(X_{R2,Gen} - X_{R3,Gen})$;
- DE/best/1: $V_{i,Gen} = X_{best,Gen} + F(X_{i1,Gen} - X_{R2,Gen})$;
- DE/rand/2: $V_{i,Gen} = X_{R1,Gen} + F(X_{R2,Gen} - X_{R3,Gen}) + F(X_{i4,Gen} - X_{R5,Gen})$;

- DE/best/2: $V_{i,Gen} = X_{best,Gen} + F(X_{R1,Gen} - X_{R2,Gen}) + F(X_{R3,Gen} - X_{R4,Gen})$;
- DE/rand/1: $V_{i,Gen} = X_{i,Gen} + F(X_{best,Gen} - X_{i,Gen}) + F(X_{R1,Gen} - X_{R2,Gen})$.

Where $R1$, $R2$, $R3$, $R4$, and $R5$ are mutually exclusive random indices in [1, NP], $X_{i,Gen}$ is the current vector, $X_{best,Gen}$ is the current best solution, and $V_{i,Gen}$ is the mutant, i.e., the result of the mutation. DE/$rand$/1 is the mutation scheme used in most multi-objective Differential Evolution variations [58].

The two most commonly used crossover operators used in Differential Evolution are the binomial and the exponential crossover. Binomial crossover is independently applied to each variable. A random number between 0 and 1 is generated and compared to CR. If it is lower than CR then the value of the current variable is derived from $V_{i,Gen}$; otherwise it is inherited from $X_{i,Gen}$. In order to have at least one variable with value derived from $V_{i,Gen}$, a random number j between 1 and NP is generated and the value of the jth variable of the target is set to the value of the jth variable from $V_{i,Gen}$.

The exponential crossover sequentially copies some variables from the mutant (or donor) vector to the target. First, an initial crossover point n in [1, NP] is randomly determined and then the number of variables to be copied (L) is determined according to the crossover rate (CR). The exponential crossover is only effective when there is dependence between neighboring variables [231]. So, most of the time the binomial crossover is the more efficient one.

The combination of DE and local search is an active research area [58]. In [211], Reynoso-Meza et al. presented an algorithm combining DE with a sequential quadratic programming local search. Jia et al. [128] proposed the hybridization of DE and a chaotic local search. A Hooke–Jeeves local search method is combined with a DE in [201]. Zhang, Chen, and Xin [252] presented a distributed DE, which also used a Hooke–Jeeves local search to fine-tune the solutions. Piotrowski proposed the hybridization of DEGL (Differential Evolution with Global and Local neighborhoods) with the Nelder–Mead (Quasi-Simplex) local search [200].

In this chapter we use the DE/$rand$/1/bin [225] variation and hybridize it with a Quasi-Simplex local search.

6.5 Quasi-Simplex Local Search

The main contribution of this work is the investigation of the combination of MOEA/D and a Quasi-Simplex local search component. The use of a local search component can improve the convergence of the solutions to MOEA/D subproblems and can also find better extreme solutions, which can contribute to the spread (diversity) of the solutions.

The behavior of the Quasi-Simplex is as follows [79]. Initially, the algorithm selects the best $N+1$ solutions of the population, where N is problem size (dimension, number of variables). Then the centroid of the simplex, which is the average point from the set of best solutions, is calculated according to Equation 6.7.

$$X_c = \left(\sum_{i=1}^{n} X_i \right) / (n+1) \tag{6.7}$$

where X_c is the centroid and X_i is i^{th} best solution of the population. As is clear in Equation 6.7, the worst solution from the set is not used in the centroid calculation.

Afterwards, several transformations are applied to modify the set of best solutions. These transformations are responsible for improving the overall quality of the set and, specially, for finding new best solutions.

In this work two sets of transformations are investigated: one proposed in [79] and the other proposed in [250].

The first set of transformations is as follows:

- **Reflection**: of the worse (X_w) and the best (X_b) individuals in relation to the centroid (Equations 6.8 and 6.9, respectively).

$$X_{rw} = X_w + (X_w - X_c) \tag{6.8}$$

$$X_{rb} = X_b + (X_c - X_b) \tag{6.9}$$

where X_{rw} is the reflection of the worse solution and X_{rb} is the reflection of the best solution;

- **Expansion**: corresponds to a reflection far from the centroid. There are two types of expansions: expansion of the worst and of the best solutions. The expansions are given in Equations 6.10 and 6.11.

$$X_{ew} = X_w + (1.0 + \alpha)(X_w - X_c) \tag{6.10}$$

$$X_{eb} = X_b + (1.0 + \alpha)(X_c - X_b) \tag{6.11}$$

where X_{ew} and X_{eb} correspond to the expansion of the worst and the best solutions, respectively, and α is a uniform random number on interval (0,1);

- **Compression**: generates points between the worst solution and the centroid (X_{cw}, Equation 6.12) or between the best solution and the centroid (X_{cb}, Equation 6.13).

$$X_{cw} = X_w - \alpha(X_w - X_c), \tag{6.12}$$

$$X_{cb} = X_b - \alpha(X_c - X_b). \tag{6.13}$$

Finally, the six best solutions found are returned by the local search and introduced on the population of MOEA/D, replacing the worst solutions.

The second set of transformations is composed by four operations: Reflection, Expansion, Outside Contraction, and Inside Contraction, given in Equations 6.14, 6.15, 6.16, and 6.17.

$$X_r = (1 + \alpha)X_c - \alpha X_p \tag{6.14}$$

$$X_e = (1 + \alpha\gamma)X_c - \alpha\gamma X_p \tag{6.15}$$

$$X_{co} = (1 + \alpha\beta)X_c - \alpha\beta X_p \tag{6.16}$$

$$X_{ci} = (1 - \beta)X_c + \beta X_p \tag{6.17}$$

where X_r, X_e, X_{co}, and X_{ci} is the Reflection, Expansion, Outside Contraction, and Inside Contraction results, respectively, α, β, and γ are parameters of the transformations and set to 1, 2, and 0.5, respectively [250].

6.6 Proposed Algorithm—MOEA/D$_{QS}$

Algorithm 6 presents the pseudocode of the proposed approach, wherein $rand \in$ U[0,1]).

The first steps of MOEA/D$_{QS}$ initialize various data structures (steps 1 to 6). The weight vectors $\boldsymbol{\lambda}^i$, $i = 1, ..., N$, representing coefficients associated with each objective, are generated using a uniform distribution. The neighborhood ($B^i = \{i_1, \cdots, i_C\}$) of weight vector $\boldsymbol{\lambda}^i$ is comprised of the indices of the C weight vectors closest to $\boldsymbol{\lambda}^i$. The initial population is randomly generated and evaluated. Each individual ($\mathbf{x}^i$) is associated with the ith weight vector. The empirical ideal point ($\mathbf{z}^*$) is initialized as the minimum value of each objective found in the initial population and the current number of evaluations ($Evaluations$) is set to N.

After the initialization steps, the algorithm enters its main loop (steps 7 to 29). The Differential Evolution operators are applied considering individuals randomly selected from *scope*. In this work, *scope* can swap from the neighborhood to the entire population (and vice versa) along the evolutionary process of MOEA/D$_{QS}$. It is composed of the indices of vectors from either

Algorithm 6 Pseudocode of MOEA/D$_{QS}$.

1: Generate N weight vectors $\boldsymbol{\lambda}^i = (\lambda_1^i, \lambda_2^i, ...\lambda_M^i), i = 1, ..., N$
2: For $i = 1, \cdots, N$, define the set of indices $B^i = \{i_1, \cdots, i_C\}$ where $\{\boldsymbol{\lambda}^{i_1}, .., \boldsymbol{\lambda}^{i_C}\}$ are the C closest weight vectors to $\boldsymbol{\lambda}^i$ (by the Euclidean distance)
3: Generate an initial population $P^0 = \{\mathbf{x}^1, \cdots, \mathbf{x}^N\}$, $\mathbf{x}^i = (x_1^i, x_2^i, ...x_n^i)$
4: Evaluate each individual in the initial population P^0 and associate $\mathbf{x}^i$ with $\boldsymbol{\lambda}^i$
5: Initialize $\mathbf{z}^* = (z_1^*, \cdots, z_M^*)$ by setting $z_j^* = min_{1 \le i \le N} f_j(\mathbf{x}^i)$
6: $Evaluations = N$
7: **repeat**
8:   **for** each individual $\mathbf{x}^i$ **do**
9:     **if** $rand < \delta$ **then**
10:       $scope = B^i$
11:     **else**
12:       $scope = \{1, \cdots, N\}$
13:     **end if**
14:     Generate a new solution $\mathbf{y}$ using DE/rand/1/bin
15:     Apply polynomial mutation to produce $\mathbf{y}'$
16:     Update $\mathbf{z}^*$, $z_j^* = min(z_j^*, f_j(\mathbf{y}'))$
17:     **for** each subproblem k randomly selected from *scope* **do**
18:       **if** $g^{te}(\mathbf{y}' \mid \boldsymbol{\lambda}^k, \mathbf{z}^*) < g^{te}(\mathbf{x}^k \mid \boldsymbol{\lambda}^k, \mathbf{z}^*)$ **then**
19:         **if** a new replacement may occur **then**
20:           Replace $\mathbf{x}^k$ by $\mathbf{y}'$ and increment n_r
21:         **end if**
22:       **end if**
23:     **end for**
24:   **end for**
25:   **if** $Evaluations$ modulo $NE == 0$ **then**
26:     Apply Quasi-Simplex
27:   **end if**
28:   $Evaluations = Evaluations + N$;
29: **until** $g >$MAX-EV

the neighborhood B^i (with probability δ) or from the entire population (with probability $1-\delta$). Applying the DE operators, a modified vector $\mathbf{y}$ is generated in step 14.

The polynomial mutation in step 15 generates $\mathbf{y}' = (y_1', \cdots, y_n')$ from $\mathbf{y}$ in the following way [61]:

$$y_d' = \begin{cases} y_d + \sigma_d.(y_d^{(U_p)} - y_d^{(L_w)}), & \text{with probability } p_m \\ y_d, & \text{with probability } 1 - p_m \end{cases} \tag{6.18}$$

with

$$\sigma_d = \begin{cases} (2 \,.\, rand)^{\frac{1}{\tau+1}} - 1, & \text{if } rand < 0.5 \\ 1 - (2 - 2 \,.\, rand)^{\frac{1}{\tau+1}}, & \text{otherwise} \end{cases} \tag{6.19}$$

where *rand* $\in$ U[0,1]. The distribution index τ and the mutation rate p_m are two DE parameters, remembering $y_d^{(L_w)}$ and $y_d^{(U_p)}$ are the lower and upper bounds of the dth decision variable, respectively.

In step 16, if the new vector $\mathbf{y}'$ has an objective value better than the value stored in the empirical ideal point, $\mathbf{z}^*$ is updated with this value.

The next steps involve the population update process (steps 17 to 23) which is based on the comparison of the fitness of individuals. In the MOEA/D framework, the fitness of an individual is measured according to a decomposition function. In this work the Tchebycheff function is used (Equation 6.4).

According to what is selected for the *scope* (steps 10 or 12), the neighborhood or the entire population is updated. The population update is as follows: if a new replacement may occur (i.e. while $n_r < NR$ and there are unselected indices in *scope*), a random index (k) from *scope* is chosen. If $\mathbf{y}'$ has a better Tchebycheff value than $\mathbf{x}^k$ (both using the kth weight vector, $\boldsymbol{\lambda}^k$) then $\mathbf{y}'$ replaces $\mathbf{x}^k$ and the number of updated vectors (n_r) is incremented. To avoid the proliferation of $\mathbf{y}'$ to a great part of the population, a maximum number of updates (NR) are used.

If the current number of evaluations (*Evaluations*) is a multiple of NE (number of evaluations between consecutive local search applications), then the Quasi-Simplex local search is executed. There are two set of transformations that can be applied during the Quasi-Simplex execution. The first set is composed of the transformations described in Equations 6.8 to 6.13 while the second set is described in Equations 6.14 to 6.17. The solutions returned by the Quasi-Simplex are compared with those in the neighborhood of the current solution. If a solution returned by the local search is better than a best solution selected in the neighborhood, then the solution of the Quasi-Simplex is inserted into the population and the worst solution is removed. This process is similar in nature to the population update (Steps 17 to 23), but operates over the set of solutions returned by the Quasi-Simplex.

The evolutionary process stops when the maximum number of evaluations (MAX-EV) is reached and MOEA/D$_{QS}$ outputs the Pareto set and Pareto front approximations.

6.7 Experiments and Results

The experimental section of this chapter is separated in two parts: (i) effect of different local search configurations and (ii) comparison with literature. The first part, which considers only the CEC 2009 Multi-Objective Competition [255] benchmark, is responsible for finding a good set of parameters for the proposed approach while the second part, which also considers three additional benchmarks (WFG, DTLZ and ZDT) is used to compare MOEA/D$_{QS}$'s performance to two algorithms from the literature (MOEA/D and NSGA-II).

The analysis of the results was based on multi-objective quality indicators [137]. During the parameters setting only the Hypervolume and additive Unary-ϵ indicators were used. In the comparison with the literature, the Inverted Generational Distance (IGD) was also applied.

All algorithms were executed, 50 times and the termination criteria for all of them was 300,000 evaluations; therefore, the comparison among them are fair.

6.7.1 Addressed Problems

The proposed approach is tested using four different MOP benchmarks: CEC 2009 [255], ZDT [263], DTLZ[72], and WFG [121].

The CEC 2009 Multi-Objective Competition benchmark [255] is composed of 10 problems with different characteristics. The problems are named UF1, UF2,..., and UF10. UF1 through UF7 are bi-objective problems while UF8, UF9, and UF10 have three objectives.

For ZDT, we use five bi-objective, continuous MOPs, named ZDT1-4, and ZDT6 [263]. The MOPs focus on different characteristics: ZDT1, ZDT2, and ZDT3 have a convex, non-convex, and disconnected PF, respectively; while ZDT4 contains many local PFs, and in ZDT6 the distribution of solutions is very non-uniform.

We use all nine WFG instances in their bi-objective formulations. WFG1 and WFG2 are convex, WFG3 is linear while WFG4–9 are concave. WFG1, WFG7–9 are biased while the other instances are unbiased. Furthermore, WFG2, WFG4, and WFG9 are multimodal while the others are unimodal and WFG1, WFG4, WFG5, and WFG7 are separable while the others are not.

For DTLZ, seven instances (DTLZ1–7) with two objectives are used. DTLZ1 has a linear Pareto optimal front. DTLZ2 is a generic sphere problem. DTLZ3 is a variation of DTLZ3 with multiple local Pareto optimal fronts. DTLZ4 is a variation of DTLZ2 that is used to evaluate if an algorithm is capable of maintaining a good distribution of solutions. DTLZ5 is a variation of DTLZ2 used to infer the convergence of the algorithms and also to be easily visualized. DTLZ6 is a more complex (more difficult to converge) version of DTLZ5. Finally, DTLZ7 is a MOP with 2^{M-1} disconnected Pareto optimal regions in the search space.

6.7.2 Quality Indicators

Quality indicators are functions that assign a real number to one or more approximation sets. Three quality indicators are applied: hypervolume, additive unary-ϵ [265], and IGD [219].

The Hypervolume indicator proposed by [260] measures the hypervolume of the objective space portion that is weakly dominated by an aproximation set A. The objective is to maximize this indicator. The calculation of hypervolume needs a (set of) reference point(s). The reference point is usually the nadir point, which is dominated by all other points. One major disadvantage of the hypervolume indicator is its computational time, which grows exponentially with the number of objectives. It should be noted that some implementations of this indicator consider the difference of the hypervolume with a reference set (R); in this case the resulting value should be minimized. In this work the first implementation is utilized.

The additive unary-ϵ indicator calculates the lowest amount, which, when added to all solutions of a reference set, makes it weakly dominated. Thus, this indicator should be minimized. This indicator is not computationally expensive.

The IGD (Inverted Generational Distance) measures the distance between the approximation front (X) and the Pareto front. Smaller distances indicate better approximations; therefore this indicator should be minimized. It can be formulated as:

$$IGD(X, Y) := \frac{1}{M}\left(\sum_{i=1}^{M} dist(y_i, X)^p\right)^{1/p} = \frac{\|d_{YX}\|_p}{M} \tag{6.20}$$

where X=x1, x2, ..., xK, Y=y1, y2, ..., yM, dist is the Euclidian distance, K is the number of solutions in the approximation front, and M is the number of solutions in the Pareto front.

6.7.3 Effect of Different Local Search Configurations

Several parameters influence the behavior and performance of the Quasi-Simplex local search in MOEA/D$_{QS}$. In this subsection we investigate the effects of different parameter configuration. The effect of the following parameters were investigated:

- The local search formulation;
- The scope of the search;
- The selection of solutions to apply the local search; and
- The number of evaluations between local search applications.

We only investigate the parameters associated with local search, using the MOEA/D default parameters in order to limit the number of parameters under investigation and to have a fair comparison with the classical MOEA/D. As the objective was not to find the best configuration, but to find a good enough one, the parameters were investigated in a sequential order.

6.7.3.1 Effect of the Local Search Formulations

Initially, different local search formulations are tested. Two Quasi-Simplex formulations are investigated: *qs* by Dong [79] and *qsC* by Coello [250] (see Section 6.5).

Table 6.1 shows the Hypervolume results for both formulations while Table 6.2 presents the Unary-ϵ results. It is clear from both tables that the version with the *qsC* formulation is better for UF1 through UF7 while the *qs* formulation is better for UF9 and UF10. For the UF8 instance, the *qsC* is better for the Hypervolume indicator while the *qs* formulation is better for the Unary-ϵ indicator. Therefore, the *qsC* is better in more instances and is chosen for the next steps of the local search parameters investigation.

TABLE 6.1: Mean and standard deviation for the hypervolume indicator of the local search formulations

	MOEA/Dqs	MOEA/DqsC
UF1	$2.22e-01_{8.3e-02}$	$3.28e-01_{5.2e-02}$
UF2	$2.44e-01_{1.5e-01}$	$5.11e-01_{1.0e-01}$
UF3	$1.77e-01_{1.2e-01}$	$4.47e-01_{7.4e-02}$
UF4	$1.76e-01_{9.8e-03}$	$2.28e-01_{1.4e-02}$
UF5	$3.31e-03_{1.7e-02}$	$1.44e-02_{4.0e-02}$
UF6	$2.08e-02_{4.8e-02}$	$5.07e-02_{7.2e-02}$
UF7	$1.33e-01_{5.8e-02}$	$2.25e-01_{9.5e-02}$
UF8	$3.06e-01_{4.6e-03}$	$3.08e-01_{1.7e-03}$
UF9	$5.94e-01_{5.0e-02}$	$5.90e-01_{4.9e-02}$
UF10	$6.10e-02_{1.6e-02}$	$3.95e-02_{2.9e-02}$

TABLE 6.2: Mean and standard deviation for the unary-ϵ indicator of the local search formulations

	MOEADqs	MOEA/DqsC
UF1	$7.26e-01_{1.2e-01}$	$5.82e-01_{1.1e-01}$
UF2	$6.71e-01_{2.4e-01}$	$2.28e-01_{2.0e-01}$
UF3	$7.10e-01_{2.0e-01}$	$2.59e-01_{1.1e-01}$
UF4	$1.49e-01_{1.3e-02}$	$1.02e-01_{1.9e-02}$
UF5	$1.08e+00_{1.2e-01}$	$1.00e+00_{1.2e-01}$
UF6	$9.77e-01_{8.2e-02}$	$9.31e-01_{1.2e-01}$
UF7	$8.31e-01_{8.5e-02}$	$6.36e-01_{2.4e-01}$
UF8	$2.57e-01_{3.4e-03}$	$2.58e-01_{1.9e-03}$
UF9	$3.29e-01_{1.3e-01}$	$3.56e-01_{1.3e-01}$
UF10	$8.10e-01_{6.5e-02}$	$9.10e-01_{1.2e-01}$

6.7.3.2 Effect of the Local Search Scope

In the next test, we investigated the local search scope parameter. The local search can choose solutions from the entire population (MOEA/DqsC) or only solutions from the neighborhood (MOEA/DqsNC) to compose the set of best solutions used on the Quasi-Simplex.

The Table 6.3 shows the Hypervolume results while Table 6.4 shows the Unary-ϵ indicator results. MOEA/DqsNC was better than *MOEA/DqsC* in all instances with the exception of UF10 with respect to the hypervolume. For the

TABLE 6.3: Mean and standard deviation for the hypervolume indicator of the lLocal search scope

	MOEA/DqsC	MOEA/DqsNC
UF1	$3.28e-01_{5.2e-02}$	$6.65e-01_{1.5e-04}$
UF2	$5.11e-01_{1.0e-01}$	$6.57e-01_{3.6e-03}$
UF3	$4.47e-01_{7.4e-02}$	$6.49e-01_{1.3e-02}$
UF4	$2.28e-01_{1.4e-02}$	$2.50e-01_{5.0e-03}$
UF5	$1.44e-02_{4.0e-02}$	$3.79e-02_{5.8e-02}$
UF6	$5.07e-02_{7.2e-02}$	$2.10e-01_{8.3e-02}$
UF7	$2.25e-01_{9.5e-02}$	$4.97e-01_{6.3e-04}$
UF8	$3.08e-01_{1.7e-03}$	$3.42e-01_{1.9e-02}$
UF9	$5.90e-01_{4.9e-02}$	$6.79e-01_{4.4e-02}$
UF10	$3.95e-02_{2.9e-02}$	$2.43e-02_{1.9e-02}$

TABLE 6.4: Mean and standard deviation for the unary-ϵ indicator of the local search scope

	MOEA/DqsC	MOEA/DqsNC
UF1	$5.82e-01_{1.1e-01}$	$4.53e-03_{1.5e-03}$
UF2	$2.28e-01_{2.0e-01}$	$5.37e-02_{1.7e-02}$
UF3	$2.59e-01_{1.1e-01}$	$4.73e-02_{4.8e-02}$
UF4	$1.02e-01_{1.9e-02}$	$7.04e-02_{5.5e-03}$
UF5	$1.00e+00_{1.2e-01}$	$5.14e-01_{1.2e-01}$
UF6	$9.31e-01_{1.2e-01}$	$2.99e-01_{1.9e-01}$
UF7	$6.36e-01_{2.4e-01}$	$2.13e-02_{5.4e-03}$
UF8	$2.58e-01_{1.9e-03}$	$2.35e-01_{2.2e-02}$
UF9	$3.56e-01_{1.3e-01}$	$1.92e-01_{1.0e-01}$
UF10	$9.10e-01_{1.2e-01}$	$8.51e-01_{6.9e-02}$

Unary-ϵ indicator, *MOEA/DqsNC* always obtained better results. Therefore, only versions considering the neighborhood scope are considered from here on.

6.7.3.3 Effect of the Selection of Solutions to Apply the Local Search

We investigated three options for the selection of solutions to apply the Quasi-Simplex local search: all solutions (*MOEA/DqsNC*), only on solutions that improved since the last application of the local search (*MOEA/DqsNCB*), and only on solutions that had not improved since the last local search application (*MOEA/DqsNCW*).

The Table 6.5 shows results for the Hypervolume quality indicator while Table 6.6 shows results for the Unary-ϵ indicator. For both indicators, *MOEA/DqsNCB* was better than *MOEA/DqsNCW* in all instances but UF2 and UF8. So, in the next experiments, we only consider versions that apply the local search to the solutions that improved since the last local search application.

6.7.3.4 Effect of the Number of Evaluations between Local Search Applications

The final Quasi-Simplex parameter that needs to be set is the number of evaluations between local search applications. If this parameter is too small,

TABLE 6.5: Mean and standard deviation for the hypervolume indicator of selection

	MOEA/DqsNC	MOEA/DqsNCB	MOEA/DqsNCW
UF1	$6.65e-01_{1.5e-04}$	$6.65e-01_{9.7e-05}$	$6.65e-01_{1.0e-04}$
UF2	$6.57e-01_{3.6e-03}$	$6.57e-01_{3.6e-03}$	$6.59e-01_{2.9e-03}$
UF3	$6.49e-01_{1.3e-02}$	$6.52e-01_{8.6e-03}$	$6.47e-01_{1.7e-02}$
UF4	$2.50e-01_{5.0e-03}$	$2.53e-01_{4.4e-03}$	$2.53e-01_{4.7e-03}$
UF5	$3.79e-02_{5.8e-02}$	$4.70e-02_{6.1e-02}$	$2.46e-02_{5.1e-02}$
UF6	$2.10e-01_{8.3e-02}$	$2.31e-01_{5.9e-02}$	$2.14e-01_{6.7e-02}$
UF7	$4.97e-01_{6.3e-04}$	$4.97e-01_{4.5e-04}$	$4.97e-01_{1.0e-03}$
UF8	$3.42e-01_{1.9e-02}$	$3.43e-01_{1.6e-02}$	$3.58e-01_{1.8e-02}$
UF9	$6.79e-01_{4.4e-02}$	$7.14e-01_{3.0e-02}$	$7.08e-01_{3.8e-02}$
UF10	$2.43e-02_{1.9e-02}$	$2.94e-02_{1.9e-02}$	$2.85e-02_{1.9e-02}$

TABLE 6.6: Mean and standard deviation for the unary-ϵ of selection

	MOEA/DqsNC	MOEA/DqsNCB	MOEA/DqsNCW
UF1	$4.53e-03_{1.5e-03}$	$3.65e-03_{1.1e-03}$	$3.88e-03_{1.7e-03}$
UF2	$5.37e-02_{1.7e-02}$	$5.46e-02_{1.6e-02}$	$4.76e-02_{1.5e-02}$
UF3	$4.73e-02_{4.8e-02}$	$3.76e-02_{3.3e-02}$	$4.39e-02_{4.0e-02}$
UF4	$7.04e-02_{5.5e-03}$	$6.70e-02_{6.5e-03}$	$6.86e-02_{7.1e-03}$
UF5	$5.14e-01_{1.2e-01}$	$4.79e-01_{8.0e-02}$	$4.94e-01_{6.0e-02}$
UF6	$2.99e-01_{1.9e-01}$	$2.56e-01_{9.7e-02}$	$2.73e-01_{1.1e-01}$
UF7	$2.13e-02_{5.4e-03}$	$1.67e-02_{4.4e-03}$	$1.83e-02_{6.6e-03}$
UF8	$2.35e-01_{2.2e-02}$	$2.41e-01_{2.0e-02}$	$2.26e-01_{2.6e-02}$
UF9	$1.92e-01_{1.0e-01}$	$1.27e-01_{7.2e-02}$	$1.39e-01_{9.3e-02}$
UF10	$8.51e-01_{6.9e-02}$	$8.49e-01_{5.7e-02}$	$8.51e-01_{6.1e-02}$

then the algorithm probably converges to local optimal solutions. Otherwise, if it is too large, then the local search have little or no influence on the results. The values investigated for this parameter are 5,000, 15,000 and 25,000 evaluations between consecutive applications.

The results for the Hypervolume and the Unary-ϵ indicators are presented in Tables 6.7 and 6.8, respectively. *MOEA/DqsNCB15000* was better for all instances in all indicators with the exceptions of UF2, UF8 (where *MOEA/DqsNCB25000* was better), and UF10 (where *MOEA/DqsNCB5000* was better).

Therefore, the best set of parameters found in this work is as follows: use the Quasi-Simplex formulation proposed in [250], the scope of the solutions considered to compose the set of best solutions considers only solutions from the neighborhood of the current solution, the local search is only applied to solutions that improved since the last application of the local search, and the number of evaluations between consecutive local search applications is set to 15,000.

6.7.4 Comparison with Literature

In this subsection, the best version found in the previous experiments (MOEA/DqsNCB15000, henceforth called MOEA/D_{QS}) is compared with the original MOEA/D and the NSGA-II algorithms.

TABLE 6.7: Mean and standard deviation for the hypervolume indicator of number of evaluations

	MOEA/DqsNCB5000	MOEA/DqsNCB15000	MOEA/DqsNCB25000
UF1	$6.65e-01_{1.9e-04}$	$6.65e-01_{9.7e-05}$	$6.65e-01_{1.3e-04}$
UF2	$6.54e-01_{5.0e-03}$	$6.57e-01_{3.6e-03}$	$6.58e-01_{2.8e-03}$
UF3	$6.41e-01_{2.0e-02}$	$6.52e-01_{8.6e-03}$	$6.45e-01_{1.7e-02}$
UF4	$2.51e-01_{4.5e-03}$	$2.53e-01_{4.4e-03}$	$2.50e-01_{4.5e-03}$
UF5	$3.80e-02_{5.4e-02}$	$4.70e-02_{6.1e-02}$	$2.13e-02_{4.6e-02}$
UF6	$2.03e-01_{8.2e-02}$	$2.31e-01_{5.9e-02}$	$2.05e-01_{8.8e-02}$
UF7	$4.96e-01_{1.0e-03}$	$4.97e-01_{4.5e-04}$	$4.97e-01_{6.6e-04}$
UF8	$2.96e-01_{2.4e-02}$	$3.43e-01_{1.6e-02}$	$3.52e-01_{1.8e-02}$
UF9	$6.23e-01_{6.1e-02}$	$7.14e-01_{3.0e-02}$	$6.96e-01_{4.0e-02}$
UF10	$3.12e-02_{2.0e-02}$	$2.94e-02_{1.9e-02}$	$2.87e-02_{1.8e-02}$

TABLE 6.8: Mean and standard deviation for the unary-ϵ indicator of number of evaluations

	MOEA/DqsNCB5000	MOEA/DqsNCB15000	MOEA/DqsNCB25000
UF1	$4.82e-03_{2.1e-03}$	$3.65e-03_{1.1e-03}$	$4.21e-03_{1.7e-03}$
UF2	$6.43e-02_{2.4e-02}$	$5.46e-02_{1.6e-02}$	$5.15e-02_{1.5e-02}$
UF3	$6.11e-02_{5.2e-02}$	$3.76e-02_{3.3e-02}$	$6.16e-02_{6.5e-02}$
UF4	$6.83e-02_{5.8e-03}$	$6.70e-02_{6.5e-03}$	$6.94e-02_{5.8e-03}$
UF5	$5.07e-01_{1.1e-01}$	$4.79e-01_{8.0e-02}$	$5.05e-01_{5.8e-02}$
UF6	$3.46e-01_{2.3e-01}$	$2.56e-01_{9.7e-02}$	$2.95e-01_{1.5e-01}$
UF7	$2.33e-02_{1.3e-02}$	$1.67e-02_{4.4e-03}$	$1.88e-02_{4.9e-03}$
UF8	$2.98e-01_{1.1e-01}$	$2.41e-01_{2.0e-02}$	$2.31e-01_{2.2e-02}$
UF9	$2.88e-01_{1.3e-01}$	$1.27e-01_{7.2e-02}$	$1.67e-01_{9.5e-02}$
UF10	$8.41e-01_{5.4e-02}$	$8.49e-01_{5.7e-02}$	$8.49e-01_{5.9e-02}$

In this experiment the CEC 2009 [255], WFG [121], DTLZ [72], and ZDT [264] are considered. The hypervolume, addictive unary-ϵ, and IGD are used as quality indicators.

6.7.4.1 Benchmark CEC 2009

Table 6.9 shows the results for three algorithms with respect to the Hypervolume indicator. MOEA/D_{QS} was the best or second best algorithm for all instances, being the best algorithm for UF1, UF3, UF7, and UF9. NSGA-II was the best algorithm for UF4, UF5, UF6, and UF10 while MOEA/D was the best algorithm for UF2 and UF8 and the second best algorithm for UF1, UF3, UF7, and UF9. Therefore, it is possible to affirm that according to the Hypervolume indicator, MOEA/D_{QS} was the best algorithm.

According to Unary-ϵ results presented in Table 6.10, MOEA/D_{QS} was the best algorithm for UF1, UF3, UF6, UF7, UF9, and UF10 and second best for UF2, UF5, and UF8. NSGA-II was the best algorithm for UF4 and UF5 while MOEA/D was the best algorithm for UF2 and UF8 and the second best algorithm for UF1, UF3, UF4, UF6, UF7, and UF9. So, it is possible to affirm that according to the Unary-ϵ indicator, MOEA/D_{QS} was the best algorithm.

Table 6.11 shows the results for the IGD indicator. MOEA/D_{QS} was the best or second best algorithm for all instances, being the best algorithm for UF1, UF3, UF6, UF7, and UF9. NSGA-II was the best algorithm for UF4,

TABLE 6.9: Mean and standard deviation for the hypervolume indicator on CEC 2009 benchmark

	MOEA/D$_{QS}$	MOEA/D	NSGA-II
UF1	$6.65e-01_{9.7e-05}$	$6.65e-01_{1.6e-04}$	$5.65e-01_{1.5e-02}$
UF2	$6.57e-01_{3.6e-03}$	$6.58e-01_{2.6e-03}$	$6.42e-01_{5.0e-03}$
UF3	$6.52e-01_{8.6e-03}$	$6.51e-01_{1.2e-02}$	$5.28e-01_{2.2e-02}$
UF4	$2.53e-01_{4.4e-03}$	$2.52e-01_{4.8e-03}$	$2.76e-01_{3.6e-04}$
UF5	$4.70e-02_{6.1e-02}$	$3.70e-02_{5.4e-02}$	$1.67e-01_{6.1e-02}$
UF6	$2.31e-01_{5.9e-02}$	$2.20e-01_{5.0e-02}$	$2.67e-01_{5.3e-02}$
UF7	$4.97e-01_{4.5e-04}$	$4.97e-01_{1.1e-03}$	$4.49e-01_{1.3e-02}$
UF8	$3.43e-01_{1.6e-02}$	$3.56e-01_{1.8e-02}$	$2.34e-01_{3.8e-02}$
UF9	$7.14e-01_{3.0e-02}$	$7.02e-01_{4.7e-02}$	$6.04e-01_{3.9e-02}$
UF10	$2.94e-02_{1.9e-02}$	$1.82e-02_{1.5e-02}$	$7.09e-02_{6.8e-02}$

TABLE 6.10: Mean and standard deviation for the unary-ϵ indicator on CEC 2009 benchmark

	MOEA/D$_{QS}$	MOEA/D	NSGA-II
UF1	$3.65e-03_{1.1e-03}$	$3.96e-03_{1.4e-03}$	$1.54e-01_{2.2e-02}$
UF2	$5.46e-02_{1.6e-02}$	$4.91e-02_{1.5e-02}$	$8.03e-02_{1.8e-02}$
UF3	$3.76e-02_{3.3e-02}$	$4.11e-02_{4.9e-02}$	$2.11e-01_{5.2e-02}$
UF4	$6.70e-02_{6.5e-03}$	$6.65e-02_{6.1e-03}$	$3.56e-02_{1.7e-03}$
UF5	$4.79e-01_{8.0e-02}$	$4.79e-01_{6.5e-02}$	$3.85e-01_{6.7e-02}$
UF6	$2.56e-01_{9.7e-02}$	$2.72e-01_{9.6e-02}$	$3.04e-01_{8.9e-02}$
UF7	$1.67e-02_{4.4e-03}$	$1.98e-02_{1.6e-02}$	$1.33e-01_{2.5e-02}$
UF8	$2.41e-01_{2.0e-02}$	$2.27e-01_{2.2e-02}$	$5.61e-01_{1.7e-01}$
UF9	$1.27e-01_{7.2e-02}$	$1.53e-01_{1.2e-01}$	$2.74e-01_{1.2e-01}$
UF10	$8.49e-01_{5.7e-02}$	$8.88e-01_{5.6e-02}$	$8.53e-01_{1.3e-01}$

UF5, and UF10 while MOEA/D was the best algorithm for UF2 and UF8 and the second best algorithm for UF1, UF3, UF6, UF7, and UF9. Therefore, it is possible to affirm that according to the IGD indicator, MOEA/D$_{QS}$ was the best algorithm. As the IGD was not used during the parameter setting stage, this quality indicator is better evidence of the superior performance of MOEA/D$_{QS}$ on CEC 2009 instances.

As *MOEA/D$_{QS}$* was the best or second best algorithm for all indicators in all instances with the exception of UF4 using the Unary-ϵ indicator, it can be considered the best algorithm among those compared.

TABLE 6.11: Mean and standard deviation for the IGD indicator on CEC 2009 benchmark

	MOEA/D$_{QS}$	MOEA/D	NSGA-II
UF1	$3.76e-05_{3.4e-06}$	$3.89e-05_{6.5e-06}$	$2.92e-03_{3.7e-04}$
UF2	$4.44e-04_{1.6e-04}$	$3.90e-04_{1.3e-04}$	$1.05e-03_{3.1e-04}$
UF3	$3.97e-04_{3.5e-04}$	$4.76e-04_{6.1e-04}$	$3.69e-03_{9.3e-04}$
UF4	$1.79e-03_{1.1e-04}$	$1.81e-03_{1.2e-04}$	$1.30e-03_{1.6e-05}$
UF5	$7.09e-02_{1.1e-02}$	$7.16e-02_{1.0e-02}$	$5.24e-02_{8.5e-03}$
UF6	$3.61e-03_{2.1e-03}$	$4.00e-03_{2.0e-03}$	$6.65e-03_{2.7e-03}$
UF7	$8.45e-05_{2.3e-05}$	$1.09e-04_{1.4e-04}$	$1.76e-03_{4.1e-04}$
UF8	$9.56e-04_{1.4e-04}$	$8.40e-04_{1.3e-04}$	$2.06e-03_{4.8e-04}$
UF9	$5.07e-04_{3.5e-04}$	$6.55e-04_{5.9e-04}$	$1.18e-03_{4.6e-04}$
UF10	$5.32e-03_{5.3e-04}$	$5.51e-03_{6.2e-04}$	$3.07e-03_{2.8e-04}$

TABLE 6.12: Mean and standard deviation for the hypervolume indicator on WFG benchmark

	MOEADqs	MOEAD	NSGA-II
WFG1	$6.26e-01_{2.7e-02}$	$6.36e-01_{1.9e-04}$	$6.35e-01_{2.7e-04}$
WFG2	$5.65e-01_{3.2e-06}$	$5.65e-01_{8.3e-06}$	$5.65e-01_{4.6e-04}$
WFG3	$4.44e-01_{2.3e-06}$	$4.44e-01_{1.6e-06}$	$4.44e-01_{2.8e-05}$
WFG4	$2.22e-01_{1.5e-05}$	$2.22e-01_{1.5e-04}$	$2.22e-01_{1.7e-05}$
WFG5	$1.99e-01_{1.1e-03}$	$2.00e-01_{3.0e-03}$	$1.99e-01_{2.2e-05}$
WFG6	$2.14e-01_{1.1e-05}$	$2.14e-01_{3.1e-06}$	$2.13e-01_{1.0e-03}$
WFG7	$2.14e-01_{3.1e-06}$	$2.14e-01_{5.2e-06}$	$2.14e-01_{1.1e-05}$
WFG8	$2.08e-01_{1.1e-02}$	$1.84e-01_{2.7e-02}$	$1.95e-01_{2.1e-02}$
WFG9	$2.45e-01_{2.1e-04}$	$2.45e-01_{7.9e-05}$	$2.44e-01_{5.9e-04}$

TABLE 6.13: Mean and standard deviation for the unary-ϵ indicator on WFG benchmark

	MOEADqs	MOEAD	NSGA-II
WFG1	$6.89e-02_{3.9e-02}$	$3.17e-02_{3.2e-03}$	$4.95e-02_{2.7e-03}$
WFG2	$4.52e-03_{3.3e-05}$	$4.60e-03_{5.0e-05}$	$1.67e-02_{9.9e-02}$
WFG3	$2.00e+00_{5.5e-06}$	$2.00e+00_{2.4e-06}$	$2.00e+00_{7.4e-05}$
WFG4	$4.49e-03_{3.9e-04}$	$7.18e-03_{1.3e-03}$	$6.72e-03_{1.3e-03}$
WFG5	$5.36e-02_{1.8e-03}$	$5.28e-02_{4.7e-03}$	$5.71e-02_{1.2e-03}$
WFG6	$3.89e-03_{1.2e-04}$	$3.87e-03_{2.3e-05}$	$6.66e-03_{1.7e-03}$
WFG7	$3.99e-03_{7.3e-05}$	$4.07e-03_{6.7e-05}$	$7.12e-03_{8.1e-04}$
WFG8	$2.47e-01_{5.0e-03}$	$2.80e-01_{8.2e-02}$	$2.53e-01_{6.0e-03}$
WFG9	$4.75e-03_{4.7e-04}$	$4.93e-03_{6.2e-04}$	$7.81e-03_{1.6e-03}$

6.7.4.2 Benchmark WFG

Table 6.12 shows the results for three algorithms with respect to the Hypervolume indicator. MOEA/D_{QS} was the best for WFG8 and second best for WFG2, WFG3, WFG4, WFG5, WFG6, WFG7, and WFG9 instances. NSGA-II was the best algorithm for WFG2, WFG4, and WFG7 while MOEA/D was the best algorithm for WFG1, WFG3, WFG5, WFG6, and WFG9. Therefore, it is possible to affirm that according to the Hypervolume indicator, MOEA/D_{QS} was the best algorithm.

According to Unary-ϵ results presented in Table 6.13, MOEA/D_{QS} was the best algorithm for WFG2, WFG4, WFG7, WFG8, and WFG9 and second best for WFG3, WFG5, and WFG6 while MOEA/D was the best algorithm for WFG1, WFG3, WFG5, and WFG6 and the second best algorithm for WFG2, WFG7, and WFG9. So, it is possible to affirm that according to the Unary-ϵ indicator, MOEA/D_{QS} was the best algorithm.

Table 6.14 shows the results for the IGD indicator. MOEA/D_{QS} was the best for WFG7 and WFG8 and second best algorithm for WFG2, WFG3, WFG4, WFG5, WFG6, and WFG9 instances. NSGA-II was the best algorithm for WFG2, WFG3, and WFG4 while MOEA/D was the best algorithm for WFG1, WFG5, WFG6, and WFG9 and the second best algorithm for WFG7. Therefore, it is possible to affirm that according to the IGD indicator MOEA/D_{QS} was the best algorithm and, therefore, the best algorithm for the WFG benchmark.

TABLE 6.14: Mean and standard deviation for the IGD indicator on WFG benchmark

	MOEADqs	MOEAD	NSGA-II
WFG1	$4.69e-04_{4.6e-04}$	$2.06e-04_{2.1e-05}$	$3.06e-04_{1.9e-05}$
WFG2	$3.48e-04_{4.9e-06}$	$3.56e-04_{6.8e-06}$	$2.14e-04_{1.1e-03}$
WFG3	$3.01e-03_{6.4e-09}$	$3.01e-03_{2.8e-09}$	$3.01e-03_{8.5e-08}$
WFG4	$3.14e-05_{5.3e-07}$	$3.71e-05_{3.3e-06}$	$2.87e-05_{1.3e-06}$
WFG5	$9.27e-04_{3.2e-05}$	$9.09e-04_{8.4e-05}$	$9.31e-04_{1.6e-07}$
WFG6	$4.55e-05_{2.5e-07}$	$4.52e-05_{6.3e-08}$	$6.99e-05_{2.7e-05}$
WFG7	$1.98e-05_{2.1e-08}$	$1.98e-05_{1.9e-08}$	$1.99e-05_{6.6e-07}$
WFG8	$8.40e-04_{3.8e-04}$	$1.88e-03_{1.1e-03}$	$1.33e-03_{7.4e-04}$
WFG9	$2.20e-05_{1.5e-06}$	$2.17e-05_{5.4e-07}$	$2.46e-05_{4.8e-06}$

TABLE 6.15: Mean and standard deviation for the hypervolume indicator on DTLZ benchmark

	MOEADqs	MOEAD	NSGA-II
DTLZ1	$8.09e-01_{2.5e-04}$	$8.08e-01_{3.3e-04}$	$8.12e-01_{7.1e-04}$
DTLZ2	$4.47e-01_{7.3e-04}$	$4.47e-01_{6.0e-04}$	$4.44e-01_{9.1e-04}$
DTLZ3	$4.47e-01_{4.8e-04}$	$4.42e-01_{3.3e-02}$	$4.43e-01_{1.9e-03}$
DTLZ4	$4.44e-01_{4.9e-04}$	$4.44e-01_{2.2e-04}$	$4.42e-01_{5.8e-04}$
DTLZ6	$9.68e-02_{9.9e-07}$	$9.68e-02_{5.7e-07}$	$9.70e-02_{5.9e-06}$
DTLZ7	$3.10e-01_{3.2e-04}$	$2.99e-01_{2.3e-02}$	$3.25e-01_{3.8e-04}$

6.7.4.3 Benchmark DTLZ

Table 6.15 shows the results for three algorithms with respect to the Hypervolume indicator. MOEA/D_{QS} was the best for DTLZ3 and second best for DTLZ1, DTLZ2, DTLZ4, and DTLZ7. NSGA-II was the best algorithm for DTLZ1, DTLZ6, and DTLZ7 while MOEA/D was the best algorithm for DTLZ2 and DTLZ4 and second best for DTLZ6. Therefore, it is possible to affirm that according to the Hypervolume indicator, MOEA/D_{QS} was the best algorithm.

According to Unary-ϵ results presented in Table 6.16, MOEA/D_{QS} was the best algorithm for any instance and second best for DTLZ1, DTLZ3, and DTLZ7 while MOEA/D was the best algorithm for DTLZ4 and the second best algorithm for DTLZ2 and DTLZ6. NSGA-II was the best algorithm for DTLZ1, DTLZ2, DTLZ3, DTLZ6, and DTLZ7. So, it is possible to affirm that according to the Unary-ϵ indicator, NSGA-II was the best algorithm.

TABLE 6.16: Mean and standard deviation for the unary-ϵ indicator on DTLZ benchmark

	MOEADqs	MOEAD	NSGA-II
DTLZ1	$2.13e-02_{5.5e-04}$	$2.15e-02_{5.6e-04}$	$1.98e-02_{1.7e-03}$
DTLZ2	$6.39e-02_{1.4e-03}$	$6.37e-02_{1.8e-03}$	$5.09e-02_{7.3e-03}$
DTLZ3	$5.95e-02_{5.0e-03}$	$7.20e-02_{7.7e-02}$	$5.02e-02_{6.1e-03}$
DTLZ4	$4.84e-02_{9.5e-03}$	$3.76e-02_{2.2e-03}$	$4.01e-02_{5.8e-03}$
DTLZ6	$6.94e-03_{2.3e-05}$	$6.94e-03_{7.2e-06}$	$1.03e-03_{2.2e-04}$
DTLZ7	$9.14e-02_{3.0e-03}$	$3.35e-01_{6.4e-01}$	$3.51e-02_{4.7e-03}$

TABLE 6.17: Mean and standard deviation for the IGD indicator on DTLZ benchmark

	MOEADqs	MOEAD	NSGA-II
DTLZ1	$1.77e-04_{7.5e-07}$	$1.77e-04_{6.6e-07}$	$1.85e-04_{4.1e-06}$
DTLZ2	$2.27e-04_{1.3e-06}$	$2.27e-04_{1.6e-06}$	$2.45e-04_{5.5e-06}$
DTLZ3	$3.80e-04_{4.0e-06}$	$4.42e-04_{4.4e-04}$	$3.94e-04_{8.8e-06}$
DTLZ4	$3.41e-04_{2.0e-05}$	$3.00e-04_{8.5e-06}$	$4.10e-04_{3.4e-05}$
DTLZ6	$1.31e-05_{2.9e-08}$	$1.31e-05_{1.8e-08}$	$5.16e-06_{3.1e-07}$
DTLZ7	$1.56e-03_{4.0e-05}$	$4.08e-03_{6.4e-03}$	$6.35e-04_{1.5e-05}$

TABLE 6.18: Mean and standard deviation for the hypervolume indicator on ZDT benchmark

	MOEADqs	MOEAD	NSGA-II
ZDT1	$6.66e-01_{1.9e-06}$	$6.65e-01_{1.2e-04}$	$6.66e-01_{1.4e-05}$
ZDT2	$3.33e-01_{5.0e-07}$	$3.32e-01_{1.1e-04}$	$3.32e-01_{1.4e-05}$
ZDT3	$5.17e-01_{2.6e-06}$	$5.17e-01_{9.7e-06}$	$5.17e-01_{4.4e-06}$
ZDT4	$6.66e-01_{9.0e-06}$	$6.66e-01_{2.1e-05}$	$6.66e-01_{2.7e-05}$
ZDT6	$4.06e-01_{2.3e-08}$	$4.06e-01_{1.5e-07}$	$4.05e-01_{2.5e-05}$

Table 6.17 shows the results for the IGD indicator. MOEA/D_{QS} was the best for DTLZ1, DTLZ2 and DTLZ3 and second best algorithm for DTLZ4 and DTLZ7 instances. NSGA-II was the best algorithm for DTLZ6 and DTLZ7 while MOEA/D was the best algorithm for DTLZ4 and the second best algorithm for DTLZ1, DTLZ2, and DTLZ6. Therefore, it is possible to affirm that according to the IGD indicator, MOEA/D_{QS} was the best algorithm.

As MOEA/D_{QS} obtained better results according to the hypervolume and IGD indicators, it can also be considered the best algorithm for the DTLZ benchmark.

6.7.4.4 Benchmark ZDT

Table 6.18 shows the results for three algorithms with respect to the Hypervolume indicator. MOEA/D_{QS} was the best for ZDT1 and ZDT2 and second best for ZDT3, ZDT4, and ZDT6. NSGA-II was the best algorithm only for ZDT3 while MOEA/D was the best algorithm for ZDT4 and ZDT6. Therefore, it is possible to affirm that according to the Hypervolume indicator, MOEA/D_{QS} was the best algorithm.

According to Unary-ϵ results presented in Table 6.19, MOEA/D_{QS} was the best algorithm for ZDT1, ZDT2, and ZDT4 instances and second best for ZDT3 and ZDT6 while MOEA/D was the best algorithm for ZDT6 and the second best algorithm for ZDT1, ZDT2, and ZDT4. NSGA-II was the best algorithm only for ZDT3. So, it is possible to affirm that according to the Unary-ϵ indicator, MOEA/D_{QS} was the best algorithm.

Table 6.20 shows the results for the IGD indicator. MOEA/D_{QS} was the best for ZDT1 and ZDT2 and second best algorithm for ZDT6 instances. NSGA-II was the best algorithm for ZDT3 and ZDT4 while MOEA/D was

TABLE 6.19: Mean and standard deviation for the unary-ϵ indicator on ZDT benchmark

	MOEADqs	MOEAD	NSGA-II
ZDT1	$1.18e-03_{1.1e-05}$	$1.47e-03_{2.9e-04}$	$2.29e-03_{2.7e-04}$
ZDT2	$1.05e-03_{2.1e-06}$	$1.24e-03_{1.5e-04}$	$2.49e-03_{5.1e-04}$
ZDT3	$2.67e-03_{4.1e-05}$	$2.67e-03_{2.6e-05}$	$1.66e-03_{3.4e-04}$
ZDT4	$1.41e-03_{6.7e-05}$	$1.41e-03_{1.1e-04}$	$2.35e-03_{4.7e-04}$
ZDT6	$8.18e-04_{2.3e-06}$	$8.17e-04_{5.3e-07}$	$2.38e-03_{3.0e-04}$

TABLE 6.20: Mean and standard deviation for the IGD indicator on ZDT benchmark

	MOEADqs	MOEAD	NSGA-II
ZDT1	$2.63e-05_{7.4e-08}$	$2.76e-05_{1.2e-06}$	$2.99e-05_{7.2e-07}$
ZDT2	$2.33e-05_{3.3e-09}$	$2.42e-05_{8.3e-07}$	$3.09e-05_{7.6e-07}$
ZDT3	$5.29e-05_{5.9e-07}$	$5.27e-05_{6.2e-07}$	$2.14e-05_{5.6e-07}$
ZDT4	$3.30e-05_{4.1e-07}$	$3.27e-05_{4.9e-07}$	$3.01e-05_{6.6e-07}$
ZDT6	$2.31e-05_{1.1e-08}$	$2.31e-05_{7.1e-09}$	$3.69e-05_{9.6e-07}$

the best algorithm for ZDT6 and the second best algorithm for ZDT1, ZDT2, ZDT3, and ZDT4. Therefore, it is possible to affirm that according to the IGD indicator, MOEA/D was the best algorithm.

Again, there was a disagreement between the results of the different indicators, but two of the three indicate that the performance of MOEA/D_{QS} was superior for the ZDT benchmark.

6.8 Final Remarks

This chapter investigated the introduction of a Quasi-Simplex local search on the $MOEA/D$ framework to improve its efficiency. Two different Quasi-Simplex formulations were considered and an analysis of the parameters associated with the local searches was done. The proposed algorithm was called MOEA/D_{QS}.

Two set of experiments were carried out: parameter configuration and comparison with literature. The parameter configuration only considered instances from the CEC 2009 Multi-Objective Competition benchmark and the hypervolume and addictive Unary-ϵ quality indicators. Therefore, the performance of the proposed algorithm in the other benchmarks in the IGD quality indicator is good evidence that a good set of parameters was found during the parameter setting stage and that this set of parameters is capable of performing adequately under different MOP characteristics.

The experiments of the comparison with the literature also consider the CEC 2009 benchmark, but also consider WFG, DTLZ, and ZDT instances. Furthermore, the IGD was also added to the pool of quality indicators used.

The proposed algorithm was compared with the original MOEA/D and the NSGA-II algorithms, two of the cited algorithms in the Evolutionary Multi-Objective literature. The results obtained show that the introduction of a local search component into the *MOEA/D* framework can increase its efficiency. Furthermore, according to the Hypervolume, Unary-ϵ, and IGD quality indicators, the proposed approach obtained better results than both *MOEA/D* and *NSGA-II*.

Future research include the investigation of other local searches, such as a Sequential Quadratic Programming variation, and the application of the proposed approach on real world problems, such as the Environmental/Economic Load Dispatch.

Chapter 7

Multi-objective Evolutionary Design of Robust Substitution Boxes

Nadia Nedjah

Department of Electronics Engineering and Telecommunications, State University of Rio de Janeiro, Brazil

Luiza de Macedo Mourelle

Department of Systems Engineering and Computation, State University of Rio de Janeiro, Brazil

7.1 Introduction

In cryptography, confusion and diffusion are two important properties of a secure cipher as identified in [220]. Confusion allows one to make the relationship between the encryption key and ciphertext as complex as possible while diffusion allows one to reduce as much as possible the dependency between the *plaintext* and the corresponding *ciphertext*. *Substitution* (a plaintext symbol is replaced by another) has been identified as a mechanism for primarily confusion. Conversely, *transposition* (rearranging the order of symbols) is a technique for diffusion. In modern cryptography, other mechanisms are used, such as linear transformations. Product ciphers use alternating substitution and transposition phases to achieve both confusion and diffusion, respectively. Here we concentrate on confusion using non-linear and non-correlated substitution boxes or simply *S-boxes.*

It is well known that the more linear and the less auto-correlated the S-box is, the more resilient is the cryptosystem that uses them. However, engineering a regular S-box that has the highest non-linearity and lowest auto-correlation properties is an *NP*-complete problem. Evolutionary computation is the ideal tool to deal with this type of problem. As there are three objectives that need to be reached, which are maximal regularity, and maximal non-linearity but also minimal auto-correlation, we propose to use multi-objective evolutionary optimization. Therefore, we exploit the game theroy [183] and more specifically the well-known Nash equilibrium strategy [184] to engineer such *resilient* substitution boxes.

Generally, the result of a cryptographic process, i.e., the *plaintext* or *ciphertext*, is required on real-time basis. Therefore, the computation performed needs to be efficiently implemented. When the time requirement is a constraint, hardware implementation of the computation is usually needed. As S-boxes are omnipresent in almost all nowadays cryptosystems, it is very interesting to have an optimized hardware implementation of the S-box used. Therefore, here we take advantage of genetic programming to yield an efficient evolvable hardware for a given S-box coding.

The rest of this chapter is organized in six sections. First, in Section 7.2, we define S-boxes more formally as well as their desirable properties. Subsequently, in Section 7.3, we present the multi-objective Nash equilibrium-based evolutionary algorithm [183], used to evolve resilient S-box coding and we give a brief description of the principles of evolvable hardware. Thereafter, in Section 7.4, we describe the S-box encoding and give the definition and implementation of the fitness evaluation of an S-box coding with respect to all three considered properties: regularity, non-linearity, and auto-correlation. In the sequel, in Section 7.5, we present the methodology we employed to evolve new compact, fast, and less demanding hardware for S-boxes and define the related fitness evaluation with respect to all three hardware characteristics: area, time, and power consumption. Then, in Section 7.6, we assess the quality of the evolved S-box codes together with the corresponding hardware. Also, we compare the characteristics of the engineered S-boxes to those used by the Data Encryption Standard (DES) [191]. Last but not least, in Section 7.7, we summarize the content of the chapter and draw some useful conclusions.

7.2 Preliminaries for Substitution Boxes

S-boxes play a basic and fundamental role in many modern block ciphers. In block ciphers, they are typically used to obscure the relationship between the plaintext and the ciphertext. Perhaps the most notorious S-boxes are those used in DES [191]. S-boxes are also used in modern cryptosystems based on AES and Kasumi. All three are called *Feistel* cryptographic algorithms [175] and have the simplified structure depicted in Figure 7.1.

An S-box can simply be seen as a Boolean function of n inputs and m outputs, often with $n > m$. Considerable research effort has been invested in designing resilient S-boxes that can resist the continuous cryptanalyst's attacks. In order to resist linear and differential cryptanalysis [28, 171], S-boxes need to be confusing or non-linear and diffusing or non auto-correlated. S-boxes also need to be non-regular. In the following, we introduce some useful formal definitions for S-box properties, which are used later in the fitness evaluation of an S-box evolved coding.

Definition 1. A *simple* S-box $\mathcal{S}$ is a Boolean function defined as $\mathcal{S}$: $\mathcal{B}^n \longmapsto \mathcal{B}$.

Definition 2. A *linear* simple S-box $\mathcal{L}$ is defined in (7.1):

$$\mathcal{L}_\beta(x) = \bigoplus_{i=0}^{i=m} \beta_i.\mathcal{L}(x_i). \tag{7.1}$$

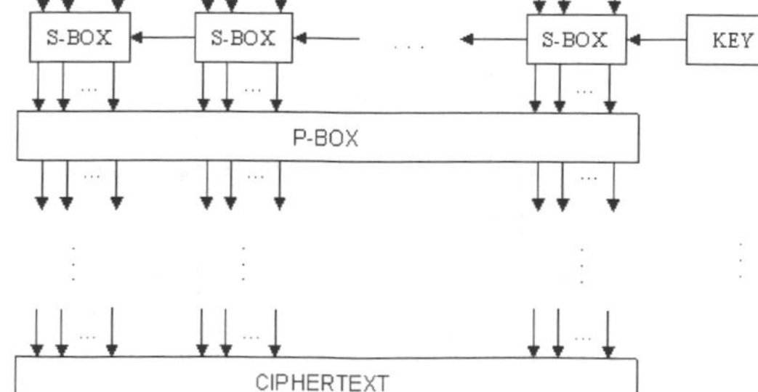

FIGURE 7.1: The simplified structure of Feistel cryptographic algorithm.

Definition 3. The *polarity* of a simple S-box $\mathcal{S}$ is defined in (7.2):

$$\hat{\mathcal{S}}(x) = (-1)^{\mathcal{S}(x)}. \tag{7.2}$$

Definition 4. The *non-correlation factor* of two simple S-boxes $\mathcal{S}$ and $\mathcal{S}'$ is defined in (7.3):

$$\mathcal{U}_{\mathcal{S},\mathcal{S}'} = \sum_{x \in \mathcal{B}^n} \hat{\mathcal{S}}(x) \times \hat{\mathcal{S}}'(x). \tag{7.3}$$

Definition 5. Two simple S-boxes $\mathcal{S}$ and $\mathcal{S}'$ are said to be non-correlated if and only $\mathcal{U}_{\mathcal{S},\mathcal{S}'} = 0$.

Definition 6. The *non-linearity* of a simple S-box $\mathcal{S}$ is measured by its non-correlation factor with all possible linear simple S-boxes defined in (7.4):

$$\mathcal{N}_{\mathcal{S}} = \frac{1}{2}\left(2^n - \max_{\alpha \in \mathcal{B}^n} |\mathcal{U}_{\mathcal{S},\mathcal{L}}|\right). \tag{7.4}$$

Definition 7. The *auto-correlation* of a simple S-box $\mathcal{S}$ is measured by its non-correlation factor with derivative S-boxes $\mathcal{D}(x) = \mathcal{S}(x) \oplus \alpha$, for all $\alpha \in \mathcal{B}^n \backslash \{0^n\}$ and defined in (7.5):

$$\mathcal{A}_{\mathcal{S}} = \max_{\alpha \in \mathcal{B}^n \backslash \{0^n\}} |\mathcal{U}_{\mathcal{S},\mathcal{D}}|. \tag{7.5}$$

Note that $\mathcal{U}_{\mathcal{S},\mathcal{D}}$ is also called *Walsh Hadamard transform* [132].

Definition 8. A simple S-box is said to be *balanced* if and only if the number of combinations $x \in \mathcal{B}^n$ such that $\mathcal{S}(x) = 0$ and the number of combinations $y \in \mathcal{B}^n$ such that $\mathcal{S}(y) = 1$ are the same. The *balance* of a simple S-box is measured using its *Hamming weight*, defined in (7.6).

$$\mathcal{W}_{\mathcal{S}} = \frac{1}{2}\left(2^n - \sum_{x \in \mathcal{B}^n} \hat{\mathcal{S}}(x)\right). \tag{7.6}$$

7.3 Evolutionary Algorithms: Nash Strategy and Evolvable Hardware

Starting from a random set of solutions, which is generally called *initial population*, an *evolutionary algorithm* breeds a population of chromosomes through a series of steps, called *generations*, using the Darwinian principle of natural selection, recombination (also called *crossover*), and *mutation*. Individuals are selected based on how much they adhere to the specified constraints. Each evolved solution is assigned a value, generally called its *fitness*, that mirrors how good it is in solving the problem in question. Evolutionary computation proceeds by first randomly creating an initial population of individuals; then, iteratively evolving to the next generation, which consists of going through two main steps, as long as the constraints are not met. The first step in a generational evolution assigns for each chromosome in the current population a fitness value that measures its adherence to the constraints while the second step creates a new population by applying the three genetic operators, which are selection, crossover, and mutation to some selected individuals. Selection is performed on the basis of the individual fitness. The fitter the individual is, the more probable it is selected to contribute to the new generational population. *Crossover* recombines two chosen solutions to create two new ones using single-point crossover, double-point crossover, or another kind of crossover operator that leads to population diversity [108]. *Mutation* yields a new individual by changing some randomly chosen genes in the selected one. The number of genes to be mutated is called *mutation degree* and how many individuals should suffer mutation is called *mutation rate.*

7.3.1 Nash Equilibrium-based Evolutionary Algorithm

This approach is inspired by the Nash strategy for economics and game theory [183, 184]. The multi-objective optimization process based on this strategy is non-cooperative in the sense that each objective is optimized separately. The basic idea consists of associating an agent or player to every objective. Each agent attempts to optimize the corresponding objective fixing the other objectives to their best values so far. As proven by Nash in [183], the Nash equilibrium point should be reached when no player can improve further the corresponding objective.

Let m be the number of objectives $f_1, \ldots, f_m$. The multi-objective genetic algorithms based on Nash strategy assigns the optimization of objective f_i to $player_i$, each of which has its own population. The process is depicted in Figure 7.2. Basically, it is a parallel genetic algorithm [80] with the exception that there are several criteria to be optimized. When a player, say $player_i$, completes an evolution generation, say t, it sends the local best solution reached f_i^t to the all $player_j$, $j \in \{1, \ldots, m\} \setminus \{i\}$, which will then be used to fix objective f_i to f_i^t during the next generation $t+1$. This evolutionary process is repeated iteratively until no player can further improve the associated criteria.

7.3.2 Evolvable Hardware

Evolutionary hardware [212] consists simply of hardware designs evolved using genetic algorithms, wherein chromosomes represent circuit designs. In general, evolutionary hardware design offers a mechanism to get a computer to provide a design of circuits without being told exactly how to do it. In short, it allows one to automatically create circuits. It does so based on a high level

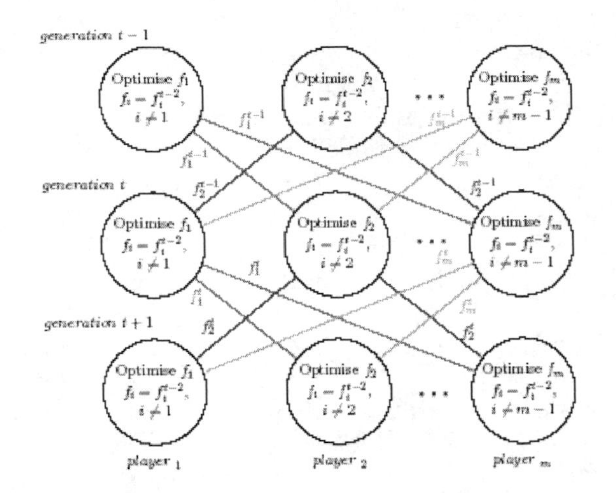

FIGURE 7.2: Multi-objective optimization using Nash strategy.

statement of the constraints the yielded circuit must obey. The input/output behavior of the expected circuit is generally considered as an omnipresent constraint. Furthermore, the generated circuit should have a minimal size.

Designing a hardware that fulfills a certain function consists of deriving from specific input/output behaviors an architecture that is operational (i.e., produces all the expected outputs from the given inputs) within a specified set of constraints. Besides the input/output behavior of the hardware, conventional designs are essentially based on knowledge and creativity, which are two human characteristics and too hard to be automated. Evolutionary hardware is a design that is generated using simulated evolution as an alternative to conventional-based electronic circuit design.

7.3.3 Crossover Operators for S-box Codings and Hardware Implementations

For both evolutionary processes (coding and hardware implementation of S-boxes), the crossover operator is implemented using four-point crossover. This is described in Figure 7.3. The four-point crossover can degenerate to either the triple, double, or single-point crossover. Moreover, these can be either horizontal or vertical. For the coding evolutionary process, the square represents the matrix of bytes while for the hardware evolution, it represents the matrix of gates and rooting. Note that this crossover can degenerate in a three-point, double-point, or single-point crossover, thus creating a better opportunity for population diversity, which leads to faster convergence.

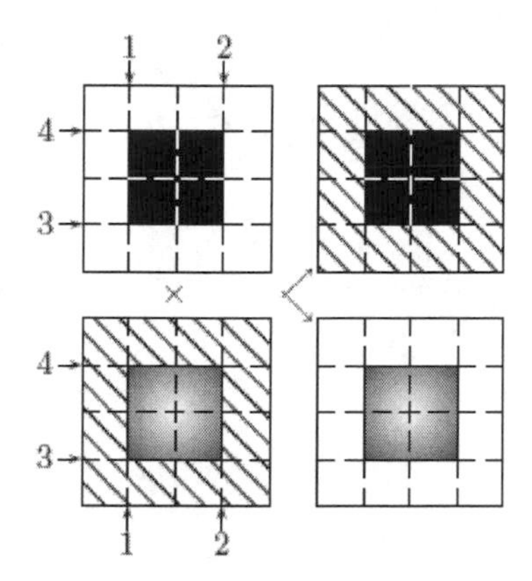

FIGURE 7.3: Four-point crossover of S-boxes.

7.4 Evolutionary Coding of Resilient S-boxes

In general, two main important concepts are crucial to any evolutionary computation: individual encoding and fitness evaluation. One needs to know how to appreciate the solutions with respect to each one of the multiple objectives. In this first evolutionary coding process, we encode an S-box simply as a matrix of bytes, which allows for an efficient application of the genetic operators: mutation and crossover. The mutation operator chooses an entry randomly and changes its value, using a fresh randomized byte. The crossover operator is handled for the evolutionary process, i.e., the S-boxes coding and hardware implementation, in the same way as described in Section 7.3.3. In the remainder of the section, we concentrate on how to evaluate the fitness of an obtained S-box. For this purpose, in the next section, we give some necessary definitions that help us implement the fitness function.

Now, let us generalize the definitions of balance, non-linearity, and autocorrelation to non-simple S-boxes, i.e., S-boxes defined as $\mathcal{S}$: $\mathcal{B}^n \longmapsto \mathcal{B}^m$.

Definition 9. An S-box $\mathcal{S}$ defined as $\mathcal{S}$: $\mathcal{B}^n \longmapsto \mathcal{B}^m$ is a concatenation of m simple S-boxes $\mathcal{S}_i$ with $1 \leq i \leq m$, such as in (7.7):

$$\mathcal{S}(x) = \mathcal{S}_1(x)\mathcal{S}_2(x)\ldots\mathcal{S}_m(x). \tag{7.7}$$

Definition 10. The *non-linearity* of S-box $\mathcal{S}$ is measured by $\mathcal{N}^*_{\mathcal{S}}$ defined in (7.8):

$$\begin{aligned} \mathcal{N}^*_{\mathcal{S}} &= \min_{\beta \in \mathcal{B}^m \setminus \{0^m\}} \mathcal{N}_{\mathcal{S}_\beta(x)}, \text{ wherein} \\ \mathcal{S}_\beta(x) &= \bigoplus_{i=0}^{i=m} \beta_i \mathcal{S}_i(x). \end{aligned} \tag{7.8}$$

Definition 11. The *auto-correlation* of S-box $\mathcal{S}$ is measured by $\mathcal{A}^*_{\mathcal{S}}$ as in (7.9):

$$\begin{aligned} \mathcal{A}^*_{\mathcal{S}} &= \max_{\beta \in \mathcal{B}^m \setminus \{0^m\}} \mathcal{A}_{\mathcal{S}_\beta(x)}, \text{ wherein} \\ \mathcal{S}_\beta(x) &= \bigoplus_{i=0}^{i=m} \beta_i \mathcal{S}_i(x). \end{aligned} \tag{7.9}$$

Definition 12. An S-box $\mathcal{S}$ is said to be *regular* if and only if for each $\omega \in \mathcal{B}^m$ there exists exactly the same number of $x \in \mathcal{B}^n$ such that $\mathcal{S}(x) = \omega$. The *regularity* of an S-box can be measured by $\mathcal{W}^*_{\mathcal{S}}$ defined in (7.10):

$$\begin{aligned} \mathcal{W}^*_{\mathcal{S}} &= \max_{\beta \in \mathcal{B}^m \setminus \{0^m\}} \left| \mathcal{W}_{\mathcal{S}_\beta(x)} \right|, \text{ wherein} \\ \mathcal{S}_\beta(x) &= \bigoplus_{i=0}^{i=m} \beta_i \mathcal{S}_i(x). \end{aligned} \tag{7.10}$$

Note that a regular S-box $\mathcal{S}$ has $\mathcal{W}^*_{\mathcal{S}} = 2^{n-1}$.

The optimization objectives consist of maximizing regularity of the S-box as well as its non-linearity while minimizing its auto-correlation. These are stated in (7.11):

$$\max_{\mathcal{S}} \mathcal{W}^*_{\mathcal{S}}, \max_{\mathcal{S}} \mathcal{N}^*_{\mathcal{S}}, \min_{\mathcal{S}} \mathcal{A}^*_{\mathcal{S}}. \tag{7.11}$$

7.5 Evolvable Hardware Implementation of S-boxes

In this section, an individual is a circuit design of an S-box, which is pre-specified using its truth table form. We encode circuit schematics using a matrix of cells that may be interconnected. A cell may or may not be involved in the circuit schematics. A cell consists of a two-input logical gate or three in case of a MUX, and a single output. A cell may draw its input signals from the output signals of gates of previous rows. The gates included in the first row draw their input signals from the circuit global ones or their complements. The circuit global output signals are the output signals of the gates in the last row of the matrix.

A circuit design is said to be *fit* if, and only if, it satisfies the imposed input/output behavior. In single objective optimization, a circuit design is considered fitter than another if it has a smaller size, shorter response, or consumes less power, depending on the optimization single objective of either size, time, or power consumption minimization, respectively. In multi-objective optimization, however, the concept of fitness is not that obvious. It is extremely rare that a single design optimizes all objectives simultaneously. Instead, there normally exist several designs that provide the same balance, compromise, or *trade-off* with respect to the problem objectives. Here, we consider three objectives: hardware area ($\mathcal{H}$), response time ($\mathcal{T}$), and power dissipation ($\mathcal{P}$). Of course, the circuit evolved needs to be fit ($\mathcal{F}$).

Objective $\mathcal{H}$ is estimated by the total number of gate-equivalents required to implement the evolved circuit and objective $\mathcal{T}$ by the maximum delay occasioned by it. Objective $\mathcal{P}$ is evaluated by approximating the switching activity of each gate and the respective fanout [178].

Let $\mathcal{C}$ be a digital circuit that uses a subset (or the complete set) of the gates given. Let $gates(\mathcal{C})$ be a function that returns the set of all gates of circuit $\mathcal{C}$ and $levels(\mathcal{C})$ be a function that returns the set of all the gates of $\mathcal{C}$ grouped by level. Notice that the number of levels of a circuit coincides with the cardinality of the set expected from function levels. On the other hand, let $B(x_i)$ be the Boolean value that the circuit $\mathcal{C}$ propagates for a row of the input Boolean matrix $X : 2^{n_{in}} \times n_{in}$ assuming that the number of input signals required for circuit $\mathcal{C}$ is n_{in}. The fitness function, which allows us to determine how much an evolved circuit adheres to the specified constraints, is given as in (7.12).

$$\min \forall \mathcal{C} \;\; (\mathcal{F}(\mathcal{C}) + \omega_1 \mathcal{H}(\mathcal{C}) + \omega_2 \mathcal{T}(\mathcal{C}) + \omega_3 \mathcal{P}(\mathcal{C}))$$

$$\begin{cases} \mathcal{F}(\mathcal{C}) = & \sum_{j=1}^{n_{out}} \sum_{i|B(x_i) \neq y_{i,j}} \xi \\ \mathcal{H}(\mathcal{C}) = & \sum_{g \in gates(\mathcal{C})} gateEquiv(g) \\ \mathcal{T}(\mathcal{C}) = & \sum_{l \in levels(\mathcal{C})} \max_{g \in l} delay(g) \\ \mathcal{P}(\mathcal{C}) = & \sum_{g \in gates(\mathcal{C})} switch(g) fanout(g) \end{cases} \tag{7.12}$$

In (7.12), $y_{i,j}$ represents the expected output value of the output signal j for input combination x_i; *nout* denotes the number of output signals that circuit $\mathcal{C}$ has. For a gate g, functions $gateEquiv$, $delay$, $switch$, and $fanout$ return the number of gate-equivalent, propagation delay, number of switches, and fanout, respectively. For each error in the evolved circuit, the individual pays a penalty ξ. Constants ω_1, ω_2, and ω_3 are the weighting coefficients that allow us to specify the importance of each one of the three objective: area, response time, and power dissipation and thus evaluate the fitness of an evolved circuit. Note that we always have $\omega_1 + \omega_2 + \omega_3 = 1$. For the implementation issue, we minimized the fitness function in (7.12) for values of $\omega_1 = 0.34$, $\omega_2 = 0.33$, and $\omega_3 = 0.33$.

7.6 Performance Results

In this section, we present some figures of performance of the two evolutionary processes described in the previous section. The evolutionary coding

TABLE 7.1: Characteristics of the best S-boxes $\mathcal{S}^+$ by Millan et al. [177], Clarck et al. [191], and our approach

input× output	Millan et al.		Clarck et al.		Our approach	
	$\mathcal{N}^*_{\mathcal{S}^+}$	$\mathcal{A}^*_{\mathcal{S}^+}$	$\mathcal{N}^*_{\mathcal{S}^+}$	$\mathcal{A}^*_{\mathcal{S}^+}$	$\mathcal{N}^*_{\mathcal{S}}$	$\mathcal{A}^*_{\mathcal{S}}$
8×2	108	56	114	32	116	34
8×3	106	64	112	40	114	42
8×4	104	72	110	48	110	42
8×5	102	72	108	56	110	56
8×6	100	80	106	64	106	62
8×7	98	80	104	72	102	70

process presented here is compared to related work from [124, 177], while the evolvable hardware implementation described earlier is compared to the designs obtained using conventional methods for the S-boxes of DES [191].

7.6.1 Performance of S-box Evolutionary Coding

The Nash algorithm [183, 184], described in Section 7.3, was implemented using multi-threading available in Java™. As we have a three-objective optimization (maximizing regularity, maximizing non-linearity, and minimizing auto-correlation), our implementation has three agents, one per objective, all three running in parallel in a computer with a Hyper-Threaded Pentium-IV processor of 3.2 GHz. We used S-boxes of different sizes (i.e. number of input and output bits) to evaluate the performance of our approach. All the fittest S-boxes' we obtained through the evolutionary process are regular. The non-linearity and the auto-correlation criteria for the best S-boxes yield are given in Table 7.1 together with those obtained in [177] and [124]. For all the benchmarks, our approach performed better producing S-boxes that are more non-linear and less auto-correlated than those presented in [177] and [124]. The best solutions were always found after at most 500 generations.

In order to compare the solutions produced by the three compared approaches, we introduce the concept of *dominance* relation between two S-boxes in a multi-objective optimisation.

Definition 13. An S-box $\mathcal{S}_1$ *dominates* another S-box $\mathcal{S}_2$, denoted by $S_1 \succ S_2$ or interchangeably solution $\mathcal{S}_2$ *is dominated by* solution $\mathcal{S}_1$ if and only if $\mathcal{S}_1$ is no worse than $\mathcal{S}_2$ with respect to all objectives and $\mathcal{S}_1$ is strictly better than $\mathcal{S}_2$ in at least one objective. Otherwise, S-box $\mathcal{S}_1$ does not dominate solution $\mathcal{S}_2$ or interchangeably $\mathcal{S}_2$ is not dominated by $\mathcal{S}_1$.

So, in light of Definition 13, we can see that all the S-boxes evolved by our approach dominate those yielded by Burnett at al. [177]. Furthermore, the 8×4, 8×5, and 8×6 S-boxes we produced dominate those generated

TABLE 7.2: Characteristics of the bitslice DES S-boxes

S-box	Conventional design			Evolutionary design		
	area	time	power	area	time	power
S1	167	2.2010	981	124	1.2880	1071
S2	149	3.8290	761	117	1.1005	981
S3	153	2.4675	992	102	1.7145	412
S4	119	1.5505	571	92	0.7660	771
S5	161	2.1170	884	126	1.2760	514
S6	162	2.2395	831	111	1.9115	959
S7	148	2.6180	716	108	1.2220	801
S8	152	2.7915	1009	137	0.9895	897

by Clarck et al. too. Nevertheless, the 8×2, 8×3, and 8×7 S-boxes are non-dominated.

7.6.2 Performance of S-box Evolvable Hardware

For comparison purposes, we evolved the S-boxes of the data encryption standard (DES) and obtained the characteristics (area, time, and power) of the evolved circuits. However, for existing work on designing hardware for DES S-boxes, we could only obtain the size in terms of gate equivalent. In Table 7.2, we give the characteristics of the S-boxes of the fastest implementation of DES known as bitslice DES [144] and, in parallel, we present the characteristics of the evolved DES S-boxes.

The parameters used in the evolutionary process were 0.9 as mutation rate, 16 as mutation degree, and a population of 100 circuits. It took us about a couple of hours to evolve the designs of DES S-Boxes S1, S2, S3, and S4. The evolvable hardware yield for S-boxes S3 and S4 are given in the appendix. However, we believe that given time, the circuit designs for the S-Boxes will be much more efficient in all of the three aspects: hardware area, response time, and power consumption (switching activity only).

The chart of Figure 7.4 relates the performance factor of the bitslice DES S-boxes versus those obtained by the evolutionary process proposed. The performance factor is the product $area \times time \times power$. It is clear that the evolutionary S-box designs are far better than those designed using conventional methods.

7.7 Final Remarks

In the first part of this chapter, we used a multi-objective evolutionary algorithm based on the concept of Nash equilibrium to evolve resilient innovative

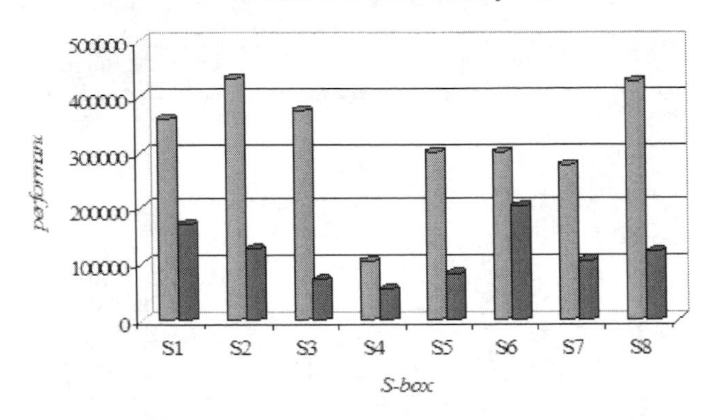

FIGURE 7.4: Performance factor of DES S-boxes: bitslice DES S-boxes vs. evolutionary S-boxes

evolutionary S-boxes. The produced S-boxes are regular in the sense of Definition 12. Moreover, considering the dominance relationship between solution S_2 in multi-objective optimization as defined in Definition 13, the generated S-boxes are better than those obtained by both Millan et al. and Clarck et al. This encourages the to pursuit of further evolution of more complex S-boxes.

In the second part of this chapter, we proposed a methodology based on evolutionary computation to automatically generate data-flow based specifications for hardware designs of S-boxes. Our aim was evolving minimal hardware specifications, i.e. hardware that minimizes the three main characteristics of a digital circuit, which are space (i.e. required gate number), time (i.e. encryption and decryption time), and power dissipation. We compared our results against the fastest existing design. The hardware evolved for the S-boxes is more efficient in terms of the required hardware area and response time, but consumes more power. Overall, however, in the trade-off between these three hardware characteristics, the evolved S-boxes are far better than the conventionally designed ones.

Acknowledgments

We are grateful to CNPq (Conselho Nacional de Desenvolvimento Científico e Tecnológico) for their continuous financial support.

Chapter 8

Multi-objective Approach to the Protein Structure Prediction Problem

Ricardo H. R. Lima

Federal University of Paraná, Brazil

Vidal Fontoura

Federal University of Paraná, Brazil

Aurora Pozo

Federal University of Paraná, Brazil

Roberto Santana

University of the Basque Country, Northern Spain

8.1 Introduction

Proteins play a fundamental task in nature, participating in many of the most important functions of living cells. These structures guarantee the correct functioning of a large number of biological entities in nature. The protein structures are the result of the so-called protein folding process in which the initially unfolded chain of amino acids is transformed into its final structure. Under suitable conditions, this structure is uniquely determined by its sequence [214]. The prediction of protein structures has a wide range of important biotechnological and medical applications, e.g., design of new proteins and folds [213, 240], structure-based drug design [204, 141], and obtaining experimental structures from incomplete nuclear magnetic resonance data [221, 207].

The determination of the final structure of a protein is a complex and challenging task even for modern super computers. This happens because it would require a huge exponential time to sample all possible configurations that a given protein sequence could adopt. Although very detailed representations of proteins exist and can be used to model the protein folding process, these representations are computationally very costly. This is why many authors as in [55, 116, 140, 152, 236] among others use simplified models to represent the protein structures. A well known model for this purpose is the *Hydrophobic–Hydrophilic* model (HP model), created by Lau and Dill [145]. Considering just two types of residues H and P in a regular lattice makes it easier to represent a protein and work with it to simulate the folding process.

Although the HP model allows a great flexibility for explaining the space of possible folds, the manipulation of a protein structure represented in the HP model requires some attention in order to respect the given problem and avoid unfeasible conformations. Another issue is the difficulty in finding good measures to verify the quality of the simplified protein conformation represented by the solution. The most common measure used for the HP model is to calculate the conformation's energy based on the number of hydrophobic contacts that exist in the fold. The question then arises of how to search for the protein configurations that optimize the energy.

Different heuristic approaches have been developed to decrease the computational complexity related to the protein structure determination process. Mono- and Multi-objective methods have been used [55, 116, 140, 152, 236, 215, 93] to find the simplified protein folds that are optimal given one or more criteria.

These approaches make use of optimization techniques like Genetic Algorithms [236], Ant Colony Optimization [222, 223], Memetic Genetic Algorithms [140], Estimation of Distribution Algorithms [215], and Multi-Objective Evolutionary Algorithms [93]. However, with few exceptions most of the previous approaches consider single-objective problem formulations and this motivated the research presented in this chapter.

This work proposes the application and comparison of a multi-objective approach to the Protein Folding Problem, considering two objectives. Using a multi-objective approach, other characteristics of the protein, and not only its energy, can be investigated. The main objective is to minimize the energy calculated from the HP model, and the second objective consists of minimizing the Euclidean distance between amino acids of the protein. The introduction of the second objective was inspired by the work of Gabriel et al. [12], in which it is mentioned that the evaluation of a structure represented by the HP model considers only the number of hydrophobic contacts, avoiding the optimization algorithms to distinguish between structures with the same number of hydrophobic contacts. Using a multi-objective approach other characteristics of the protein, and not only its energy, can be investigated. In particular, in this chapter we investigate the distance because more compact structures tend to have more hydrophobic contacts: as the lower the Euclidean distance between the amino acids is, the more compact the whole conformation will be.

Two Pareto based multi-objective evolutionary algorithms (MOEAs), NSGA-II [69] and IBEA [266], were used, because they are well known MOEAs [47] that use effective mechanisms to guarantee a good diversity of the Pareto front approximation, e.g., crowding distance mechanism and sophisticated measures to evaluate the quality of the solutions from a multi-objective point of view. Also their success when applied on the domain of other problems motivated their use in the context of this chapter.

Two versions of each algorithm were evaluated: one with the algorithms as they were originally specified and another with modifications in the initialization and mating process. A backtrack strategy is used to generate the initial population to avoid the generation of many invalid solutions. Therefore, the MOEAs will spend less time processing invalid solutions. The mating process is an important step in evolutionary algorithms since it is responsible to properly explore the search space applying the crossover and mutation operators. In multi-modal search spaces, with a lot of local optima, it makes sense to have sufficient operators that search either for higher quality or diversified solutions, depending on the region of the search space that a given EA might be stuck in. Therefore, providing evolutionary algorithms with both kinds of operators aims to better explore the search space in order to avoid local optima spots and also to exploit valleys of the search space. The mentioned modifications were introduced because the initial experiments showed that the standard MOEAs were not able to achieve satisfactory results. Another motivation for this chapter was the use of Pareto based evolutionary algorithms in order to explore the performance of this type of algorithm when applied to the PSP problem.

The remainder of this chapter is organized as follows: in the next section we briefly introduce the main aspects of the Protein Folding Problem and a review of the related works is also presented. Section 8.3 presents an overview of the Multi-Objective optimization context and the NSGA-II and IBEA algorithms are presented. Thereafter, in Section 8.4 the main contribution made in this chapter is introduced. In Section 8.5, the experimental benchmark and numerical results of the conducted experiments are presented. Finally, in Section 8.6, the conclusions of the research are given, and further work is discussed.

8.2 Protein Structure Prediction

Proteins are macromolecules made out of twenty different amino acids, also referred to as residues. An amino acid has a peptide backbone and a distinctive side chain group. The peptide bond is defined by an amino group and a carboxyl group connected to an alpha carbon to which a hydrogen and side chain group are attached.

Amino acids are combined to form sequences which are considered the primary structure of the peptides or proteins. The secondary structure is the locally ordered structure brought via hydrogen bounding mainly within the peptide backbone. The most common secondary structure elements in proteins are the alpha helix and the beta sheet. The tertiary structure is the global folding of a single polypeptide chain.

Under specific conditions, the protein sequence folds into a unique native 3D structure. Each possible protein fold has an associated energy. The *thermodynamic hypothesis* states that the native structure of a protein is the one for which the free energy achieves the global minimum. Based on this hypothesis, many methods [55, 116, 140, 152, 236] that search for the protein native structure define an approximation of the protein energy and use optimization algorithms that look for the protein fold that minimizes this energy. These approaches mainly differ in the type of energy approximation employed and in the characteristics of the protein modeling.

8.2.1 The HP Model

The protein structures are very complex. Detailed representations of proteins exist and can be used to model the protein folding; these representations are computationally very costly. Having this in mind, Lau and Dill [145] created a model called *Hydrophobic–Hydrophilic* Model (HP Model), to represent the proteins using simplifications. The model can be used either to represent proteins in a 2D space or 3D space.

The HP model considers two types of residues: hydrophobic (H) residues and hydrophilic or polar (P) residues. A protein is considered a sequence of these two types of residues, which are located in regular lattice models forming self-avoided paths. Given a pair of residues, they are considered neighbors if they are adjacent either in the chain (connected neighbors) or in the lattice but not connected in the chain (topological neighbors).

The total number of topological neighboring positions in the lattice (z) is called the lattice coordination number.

For the HP model, an energy function that measures the interaction between topological neighbor residues is defined as $\epsilon_{HH} = -1$ and $\epsilon_{HP} = \epsilon_{PP} = 0$. The HP problem consists of finding the solution that minimizes the total energy. In the linear representation of the sequence, hydrophobic residues are

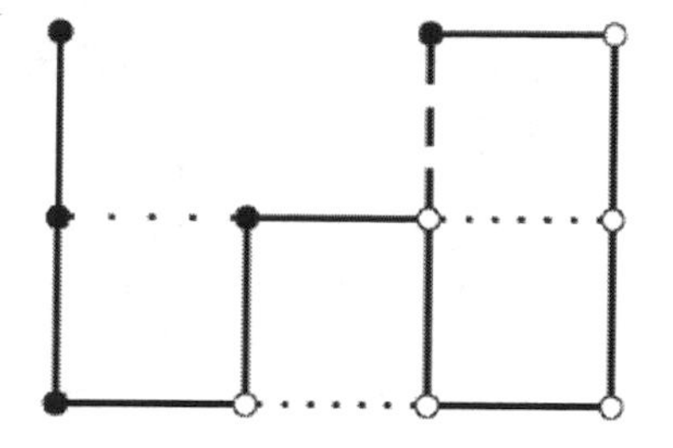

FIGURE 8.1: One possible configuration of sequence $HHHPHPPPPPH$ in the HP model. There is one HH (represented by a dotted line with wide spaces), one HP (represented by a dashed line), and two PP (represented by dotted lines) contacts.

represented with the letter H and polar ones with P. In the graphical representation, hydrophobic proteins are represented by black beads and polar proteins by white beads. Figure 8.1 shows the graphical representation of a possible configuration for the sequence $HHHPHPPPPPH$ in a 2D space. The energy that the HP model associates with this configuration is -1 because there is only one HH contact, located between the second and fifth residues.

Among many works related to the Protein Folding Problem, here are some examples of the approaches that have been used to solve it.

Unger and Moult [236] described a genetic algorithm (GA) that uses heuristic-based operators for crossover and mutation for the HP model. The algorithm outperformed many variants of Monte Carlo methods for different instances. Although good results were obtained, the GA was unable to find the optimal solution for the longest instances considered.

The multimeme algorithm (MMA) proposed by [140] is a GA combined with a set of local search methods. The algorithm, for each different instance or individual in the population, selects the local search method that best fits. Originally used to find solutions for the functional model protein, the algorithm was later improved with fuzzy logic-based local searches, leading the algorithm to produce improved results in the PSP problem.

In [116], the author uses a Chain growth algorithm, called pruned-enriched Rosenbluth method (PERM), that is based on growing the sequence conformation by adding individual particles aiming to increase good configurations and eliminating bad ones.

The Ant Colony Optimization (ACO) was also applied to the PSP problem using the HP-2D model in [222, 223]. This approach utilizes artificial ants in order to build conformations for a given HP protein sequence. A local search step is also applied to improve the results and maintain the quality of the solutions.

The work of [215] describes the use of Estimation of Distribution Algorithms (EDAs) as an efficient evolutionary algorithm that can learn and exploit the search space regularities in the form of probabilistic dependencies. In the paper new ideas about the application of EDAs to 2D and 3D simplified protein folding problems were developed. The relationship between this proposal and other population-based approaches for the protein folding problem was analyzed. The obtained results showed that EDAs can obtain superior results compared with other well-known population-based optimization algorithms for this problem.

Gabriel et al. [93] propose the use of a table-based multi-objective evolutionary algorithm initially introduced by [75], using the HP-3D model for the representation and solution evaluation. The authors also propose the use of a second objective that aims to measure the distance between hydrophobic amino acids, allowing the algorithm to distinguish between different solutions with the same energy value.

The present chapter proposes the use of two well-known MOEAs that have achieved good results when applied to other domains of problems. The main difference between this chapter and the related works [116, 140, 222, 223, 236, 215] is the multi-objective formulation. The work of Gabriel et al. [93] inspired the addition of a second objective in our approach. Although similar, the multi-objective formulation we propose is simpler, because it only considers the maximum distance between residues whereas [93] also considers the average distance between residues. The use of a backtrack initialization strategy and the modification of the mating process are also differences from the previous works. Those modifications were explored in order to improve the results in relation with the standard versions of the MOEAs. Another remarkable difference are the Pareto-based algorithms NSGA-II and IBEA that were used in this chapter whilst Gabriel et al. [93] used a different MOEA.

8.3 Multi-objective Optimization

An Evolutionary Algorithm (EA) is an optimization and search technique, highly parallel, inspired by the Darwinian principle of natural selection and genetic reproduction. The nature principles that inspire the EAs are simple. According to the theory of Charles Darwin, the principle of natural selection favors individuals with high fitness; therefore, they have high probability of reproduction. Individuals with more descendants have more chance to perpetuate their genetic code in future generations. The genetic code is what gives the identity of each individual and is represented in the chromosomes. These principles are used in the construction of computational algorithms, which search for better solutions given a specific problem by the evolution of a population of solutions encoded in artificial chromosomes—data structures used to represent a feasible solution for a given problem in the algorithm execution [193].

In general, real-world optimization problems have multiple objectives to minimize/maximize and are present in many areas of expertise. To optimize multi-objective problems, two or more objectives are considered which are usually conflicting. For these problems it is impossible to find one unique solution. A set of solutions is reached evaluating the Pareto dominance relation [17] between the solutions. The main goal is to find the solutions that are non-dominated by any other. A solution dominates others, if and only if, it was better in at least one of the objectives, without being worst in any of the objectives. The set of non-dominated solutions constitutes the Pareto front. Finding the real Pareto front is an NP-hard problem [92]; this way, the goal is to find a good approximation to this front.

Multi-Objective Evolutionary Algorithms (MOEAS) are extensions of EAs for multi-objective problems that apply the concepts of Pareto dominance to create different strategies to evolve and diversify the solutions. In this work two MOEAs were used: NSGA-II [69] and IBEA [266].

8.3.1 Non-dominated Sorting Genetic Algorithm II

The main characteristic of NSGA-II is the application of a strong elitism mechanism that at each generation sets every solution in different fronts according to the non-dominance relation.

Algorithm 7 presents the pseudocode of NSGA-II. It receives as inputs a parameter N for the population size and T as maximum number of evaluations. It starts by creating a population of size N called P_0. Then P_0 is classified according to its calculated fitness and the Non-Dominated-Sort mechanism. The classified P_0 is then submitted to a binary tournament operator to select the solutions called parents that will be used to generate the offspring. The parent solutions pass through the crossover and mutation operators generating new solutions called children. At the end of this process the offspring solutions are evaluated and put in a population called Q_0.

After this first step, P_0 and Q_0 are put together and called as an auxiliary population R. Through the non-dominated-sort, R is sorted creating the *fronts*, where solutions from the first *front* are non-dominated by any other solution, and solutions from the second front are dominated only by the solutions of the first front, and so on. For each *front*, its individuals are evaluated by the Crowding–Distance mechanism and those with higher values are stored in the next-generation population called P_t where t is the current evaluation.

After creating and filling P_t with the non-dominated solutions from all *fronts*, the whole P_t has its fitness calculated and then passes through a new process of Binary Tournament, and Crossover, and Mutation, starting a new cycle in the algorithm.

At the end, after the stop criterion is reached, the algorithm returns a set of non-dominated solutions.

Algorithm 7 NSGA-II

```
1: N ← Population Size
2: T ← Max evaluations
3: P_0 ← CreatePopulation(N);
4: CalculateFitness(P_0);
5: FastNonDominatedSort(P_0);
6: Q_0 ← 0
7: while Q_0 < N do
8:    Parents ← BinaryTournament(P_0);
9:    Offspring ← CrossoverMutation(Parents);
10:   Q_0 ← Offspring
11: end while
12: CalculateFitness(Q_0);
13: t ← 0
14: while t < T do
15:   R_t ← P_t ∪ Q_t;
16:   Fronts ← FastNonDominatedSort(R_t);
17:   P_{t+1} ← 0
18:   i ← 0
19:   while P_{t+1} + Front_i < N do
20:      CrowdingDistanceAssignment(Front_i);
21:      P_{t+1} ← P_{t+1} ∪ Front_i
22:      i ← i + 1
23:   end while
24:   CrowdingDistanceSort(Front_i);
25:   P_{t+1} ← P_{t+1} ∪ Front_i[1 : (N − P_{t+1})]
26:   Parents ← BinaryTournament(P_{t+1});
27:   Q_{t+1} ← CrossoverMutation(Parents);
28:   t ← t + 1
29: end while
30:
31: return P ← Set of non-dominated solutions.
```

8.3.2 IBEA (Indicator-Based Evolutionary Algorithm)

In the multi-objective context, optimizing consists in finding a front with good approximation to the *True Pareto front*. However, there is no general definition about what a "good approximation" of the *True* Pareto front is. Therefore, indicators have been used to evaluate the quality of a Pareto front approximation. The *hypervolume* is an example of indicator used for the evaluation and comparison of Pareto front approximations.

In IBEA, quality indicators are used to evaluate the non-dominated set of solutions [91]. To use IBEA, it is necessary to define which indicator will be used to associate each ordered pair of solutions to a scalar value. One of the most used indicators is the *hypervolume*[1] due to its capacity of evaluating the convergence and diversity of the search process at the same time [123].

$$F(x_i) = \sum_{x_j \in (P - x_i)} -e^{\frac{-I_{Hy}(x_j, x_i)}{k}}. \tag{8.1}$$

The IBEA fitness equation is given by Equation 8.1 and is used to calculate the contribution of a given solution to the indicator value of a population, where k is a scaling factor depending on I_{Hy}. The I_{Hy} represents the quality indicator to the underlying problem, being greater than 0, its commonly used with a value of 0.05. The value for $F(x_i)$ corresponds to a quality loss measure of the approximation to the Pareto front if the solution x_i was removed from the population [91], based on the value of I_{Hy}, in this case, the *hypervolume.*

Algorithm 8 receives as parameters the population size N, maximum number of evaluations T, and scale factor k. It starts by creating a population P of size N. Then it repeats the following process until the stop criterion is satisfied: through a Binary Tournament the parents are selected to be used in the Crossover and Mutation operators to generate the offspring and add them to a auxiliary population $\overline{P}$. After the reproduction step, $\overline{P}$ is added to P. While the size of P exceeds N, the worst individual evaluated by the selected indicator is removed from the population, then the population fitness is recalculated. When the algorithm stops, it returns a set of non-dominated solutions found.

8.4 A Bi-objective Optimization Approach to HP Protein Folding

Two multi-objective approaches were designed in this chapter, using the MOEAs (NSGA-II and IBEA) described in Subsection 8.3. The relative representation was chosen to represent the chromosomes. Integer vectors are used whereas the genes specify in which direction, relative to the previous residue, the next residue should be placed. The genes can assume only tree values (0,1,2): 0 indicates that the next residue should be placed on the right of the previous one, 1 indicates that the next residue should be placed in front of the previous one, and 2 indicates that the next residue should be placed on the left from the previous. Figure 8.2 shows an example of a hypothetical chromosome and the path generated by it in the 2D lattice. The first and

---

[1] *Hypervolume*: Proposed quality indicators used in the study of [269], denoted as the "size of the covered search space". This indicator has two important advantages in relation to others [261]: 1, sensitive to any kind of improvement in the approximation set in relation to other set; 2, as result of 1, the indicator guarantee that for any approximation set A that has high values of hypervolume, also has all the solutions of the true Pareto front.

Algorithm 8 IBEA

```
1: N ← Population Size
2: N̄ ← AuxiliaryPopulationSize
3: T ← Max Evaluations
4: k ← Scale factor of Fitness
5: P ← CreatePopulation(N);
6: P̄ ← CreateEmptyAuxiliaryPopulation(N̄);
7: m ← 0
8: CalculateFitness(P);
9: while m ≥ T or other stop criterion is not reached do
10:    P̄ ← BinaryTournament(P);
11:    P̄ ← CrossoverMutation(P̄);
12:    P ← P ∪ P̄
13:    m ← m + 1
14:    while Size(P) > N do
15:       x* ← WorstIndividualByFitness();
16:       RemoveFromPopulation(x*, P);
17:       CalculateFitness(P);
18:    end while
19: end while
20:
21: return P ← Set of non-dominated solutions
```

second amino acids were fixed in positions (2,3) and (3,3), respectively. The third amino acid was placed in (4,3), because the first chromosome gene is 1, and indicates that the amino acid should be placed in front of the previous. The second chromosome gene is 0, which indicates that the next amino acid should be placed on the right (4,4) of the previous. The fifth amino acid was placed in (5,4) because the chromosome gene is 2 and indicates to place the next residue on the right of the previous. The de-codification of the chromosome continues until all amino acids are placed. Note that chromosome size is always the chain length 2 because the two first amino acids are fixed.

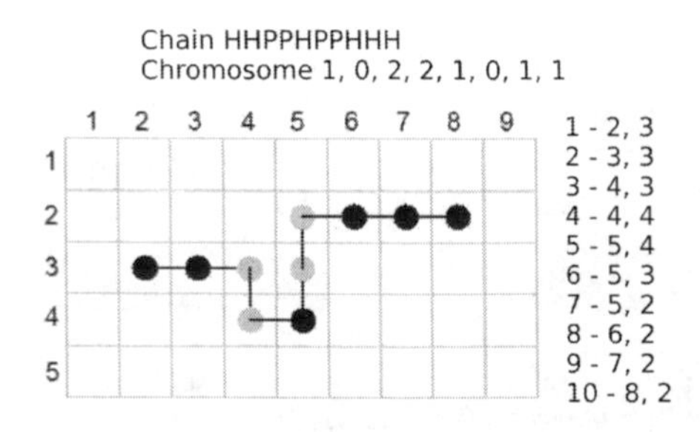

FIGURE 8.2: Example of a conformation generated by a chromosome with relative representation.

The first approach consisted on applying two well-known state-of-the-art MOEAs (IBEA and NSGA-II) to the PSP using the HP-2D model. The genetic operators used by IBEA and NSGA-II algorithms, in this approach, were the single point crossover and bit flip mutation. This combination of operators presented the best results in preliminary experiments for the PSP problem and the HP-2D model.

In the case of the second approach the IBEA and NSGA-II algorithms were modified in order to improve their results when compared with the first approach. The modifications implemented will be described next:

- **Pool of operators**: The use of traditional operators usually does not guide the search to prominent regions of the search space of the HP-2D model. In order to improve the MOEAs, a pool of operators was designed based on the literature. For every mating the crossover and mutation operators are selected randomly from the pool of operators and then applied. Also the crossover and mutation operators are always applied. This is different from the first approach which uses a probability to apply the operator. The pool of operators will be described next:
  - Single Point Crossover (1x): A single point on both parent individuals is selected. All data beyond that point in either individual will be swapped between the two parent individuals, producing two distinct offspring [114].
  - Two Point Crossover (2X): Two points are selected on both parent individuals. Everything between the two points is swapped between the parents, building two new distinct individuals [114].
  - Multi-Point Crossover (MPX): The MPX operator is similar to 2X, but the number of points, c, is a function of the sequence length, n, given by $c = int(n \times 0.1)$ [55]. The MPX operator is usually used to promote structural diversity by performing a random shuffle between individuals, although not as thoroughly as a uniform crossover.
  - Bit Flip Mutation (BFM): The BFM operator selects one random gene from a parent individual and changes it to another value, resulting in one new individual [114].
  - Local Move Mutation (LMM): The LMM operator swaps the directions between two randomly chosen consecutive genes. This operator introduces a corner movement [18]. Figure 8.3 presents an example of application of this operator.
  - Loop Move Mutation (LOMM): This operator is similar to LMM; however, it exchanges directions between genes that are five positions apart on the sequence creating a loop movement. Both LMM and LOMM are useful to generate modifications on compact structures [56].

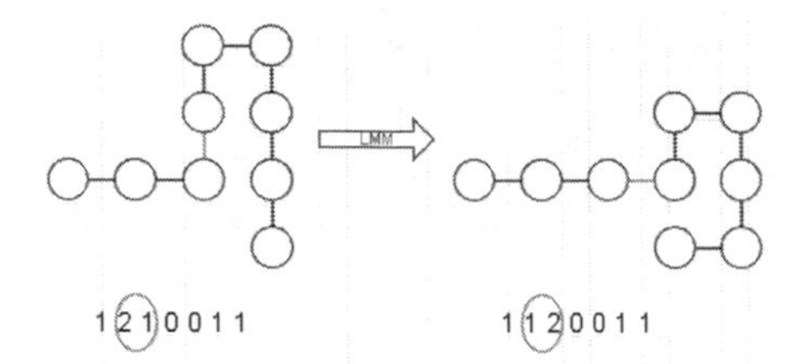

FIGURE 8.3: Example of application of the LMM operator. The genes from the red circle of left figure were swapped resulting in the right figure.

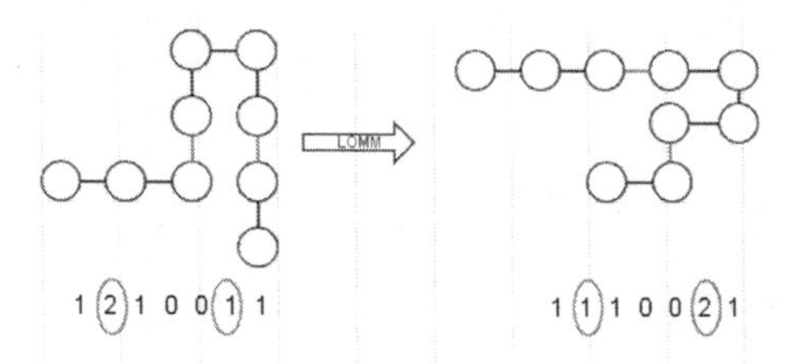

FIGURE 8.4: Example of application of the LOMM operator. The genes from the red circle of the left figure were swapped resulting in the right figure.

- Segment Mutation (SM): This operator changes a random number of consecutive genes (from two to seven) into new random directions. This operator introduces large conformational changes and has a high probability of creating collisions; in order to avoid too many invalid solutions the repair mechanism is applied on the generated solution [56]. Figure 8.5 shows an application of this operator.

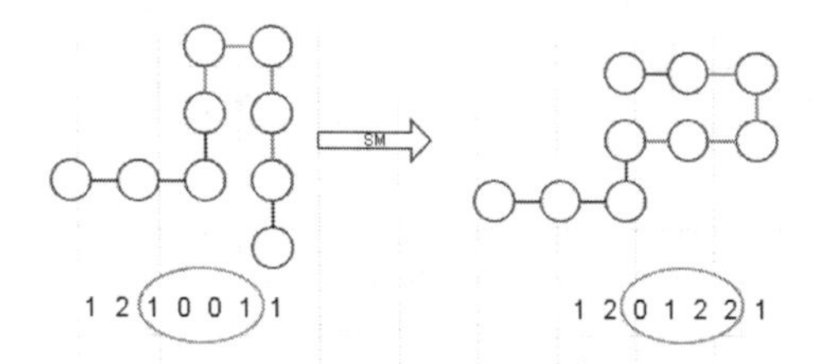

FIGURE 8.5: Example of application of the SM operator. The genes from the red circle of the left figure were swapped by random genes resulting in the right figure.

- Opposite Mutation (OM): This operator changes a random number of consecutive genes to its inverse position. In the case of the relative representation to the HP-2D model, only left and right directions can be mapped to its inverse. Figure 8.6 presents an example of application of this operator.

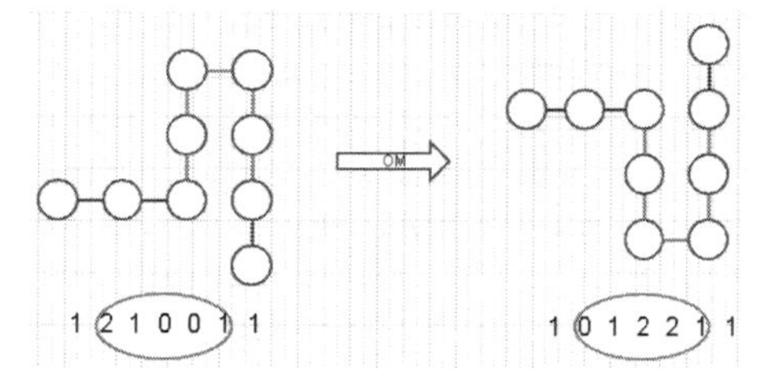

FIGURE 8.6: Example of application of the OM operator. The genes from the red circle of the left figure were swapped by random genes resulting in the right figure.

- **Backtrack Initialization**: Traditionally, the initial populations of NSGA-II and IBEA are generated randomly. The random based generation of the solutions has a great potential of generating a large number of invalid solutions for the PSP problem within the HP-2D model. Solutions that are not self-avoiding walks (SAWs) are said to contain collisions. If the initial population is fully generated randomly the evolutionary algorithms will spend time evaluating invalid solutions. In order to avoid this problem, a backtrack strategy should be applied [140]. In this approach, 20 percent of the initial population was generated using the backtrack initialization.

For both approaches the following objective functions were used:

- **Energy value**: This is the main objective and consists in the energy of given protein conformation. The goal is to minimize the energy value and it is calculated as described in Section 8.2.1. This objective guides the search progress towards regions where the energies associated to the protein conformations are minimal, thus achieving protein conformations which are closer to the native structure of a protein.

- **Minimize the distance between the two farthest residues**: This is a secondary objective and it was inspired by work presented in [93]. The motivation for this objective is that more compact conformations tend to have more hydrophobic contacts which means a lesser energy value. The distance between two residues is calculated using the Euclidean distance.

The relative representation allows the generation of invalid solutions. A solution is said to be invalid when it does not perform a SAW as mentioned before. In other words, an invalid solution is when a given residue collides

Algorithm 9 Mechanism to repair infeasible solutions

1. The position that the next residue should be placed is obtained.
2. Verification if the direction selected will cause collision.
3. If a collision is detected, a new direction is used.
4. Repeat the steps 2 and 3, until it is possible to place the next residue, or until all directions have been tested and cause collisions.
5. If it was possible to place the next residue, the mechanism achieved success; if not, the solution is considered infeasible and it will be penalized in the evaluation process.

with another already placed on the lattice. A simple mechanism for repairing these solutions was developed, and the code can be seen in the Algorithm 9.

This mechanism was implemented because in previous experiments it was observed that the number of infeasible solutions was very high. It is necessary to mention that even with the mechanism to repair solutions, there are still infeasible solutions because the mechanism cannot always repair them. Thus, infeasible solutions are penalized by adding the number of collisions to the energy value. This mechanism is used by the evaluation process of both approaches described before (MOEAs without any modifications and the MOEAs supported by the backtracking initialization and the pool of operators).

To evaluate and compare the performance of multi-objective algorithms, quality indicators are commonly used. In this study the hypervolume indicator was used, which considers the volume of the search space dominated by the known Pareto front [270] of an algorithm. Higher hypervolume value means that the quality of an algorithm is better than one with a lower hypervolume value.

Both approaches were implemented using the open source architecture from jMetal framework [81]. jMetal is easy to extend, has a well-organized structure, and also an active community.

8.5 Experiments

This section presents the set of experiments designed to evaluate the performance of the approaches introduced in Section 8.4. First, we present the HP instances used in our experiments. Then, the parameters of the MOEAs are described. We then present a comparison of the results achieved by the different MOEA variants. Finally, we present a comparison with previous results achieved by single-objective optimizers.

The HP sequences used in the experiments are shown in Table 8.1. These instances have been used in previous works such as [15, 222, 236, 54, 214,

TABLE 8.1: HP instances used in the experiments. The search space of each instance is 2^n where n is the size of the instance

Inst.	Size	$H(\mathbf{x}^*)$	Sequence
$sq1$	20	-9	$HPHPPHHPHHPHPHHPPHPH$
$sq2$	24	-9	$HHPPHPPHPPHPPHPPHPPHPPHH$
$sq3$	25	-8	$PPHPPHHP^4HHP^4HHP^4HH$
$sq4$	36	-14	$P^3HHPPHHP^5H^7PPHHP^4HHPPHPP$
$sq5$	48	-23	$PPHPPHHPPHHP^5H^{10}P^6$ $HHPPHHPPHPPH^5$
$sq6$	50	-21	$HHPHPHPHPH^4PHP^3HP^3HP^4$ $HP^3HP^3HPH^4\{PH\}^4H$
$sq7$	60	-36	$PPH^3PH^8P^3H^{10}PHP^3$ $H^{12}P^4H^6PHHPHP$
$sq8$	64	-42	$H^{12}PHPH\{PPHH\}^2PPH\{PPHH\}^2$ $PPH\{PPHH\}^2PPHPHPH^{12}$

223, 148]. The values presented in Table 8.1 correspond to the sequence identifier, the size of amino acid sequence, the best known solutions ($H(x*)$) for the HP-2D model, and the sequence itself. It is worthwhile to mention that the sequences used in this chapter were randomly generated. Hence they do not fold to a single conformation, as natural proteins, because they are not products of natural selection [41].

The configurations used for the MOEAs were defined based on the sequence length. For smaller sequences a smaller population size was used and for larger sequences a larger population size was used. The same is true in the case of the stop condition (max evaluations). Table 8.2 presents the population size and maximum number of evaluations used for each amino acid sequence. In the case of the first approach, the probability of crossover/mutation occurrence was fixed, for all sequences, to 0.9 and 0.01, respectively. The second approach does not use a probability since the operators are always applied to generate new individuals. The auxiliary population size used by the IBEA algorithm was fixed in 200 for all sequences. For each sequence the algorithms were executed 30 independent times.

8.5.1 Comparison between the Modified and Traditional Versions of the MOEAs

As mentioned in Section 8.4 the hypervolume indicator was used in order to compare the MOEA's performance. The hypervolume results are presented in Table 8.3. The hypervolume average and standard deviation, of 30 independent executions, are presented. The average values highlighted with a bold font are the highest values.

TABLE 8.2: Population size and maximum number of evaluations' configurations for each sequence

Sequences	Size	Population Size	Max Evaluations
sq1	20	100	25000
sq2	24	100	25000
sq3	25	500	250000
sq4	36	500	250000
sq5	48	1000	2500000
sq6	50	1000	2500000
sq7	60	2500	2500000
sq8	64	2500	2500000

TABLE 8.3: Results of hypervolume average/standard deviation of the MOEAs

Instance	Hypervolume Average (Std D)			
	NSGA-II	M_NSGA-II	IBEA	M_IBEA
sq1	0.742827 (0.106315)	0.720864 (0.131351)	**0.789712** (0.067660)	0.786571 (0.099424)
sq2	0.680572 (0.083445)	0.712275 (0.137226)	0.719960 (0.080727)	**0.737086** (0.095299)
sq3	0.671171 (0.129417)	0.709898 (0.124201)	0.716438 (0.148112)	**0.738017** (0.155638)
sq4	0.702280 (0689832)	0.740153 (0.075271)	0.751755 (0.092427)	**0.785728** (0.055607)
sq5	0.707654 (0.082611)	0.758128 (0.062315)	0.733464 (0.128757)	**0.807637** (0.039620)
sq6	0.667771 (0.132218)	0.774017 (0.063231)	0.728699 (0.080679)	**0.821177** (0.048124)
sq7	0.784483 (0.063257)	0.792843 (0.033062)	0.801778 (0.067111)	**0.810351** (0.054576)
sq8	0.677464 (0.041287)	0.705798 (0.053048)	0.7450656 (0.036454)	**0.811439** (0.050087)

Looking to Table 8.3 it is possible to notice that, except for *sq*1, for all sequences the M_IBEA (modified version of the IBEA with backtrack and pool of operators) obtained a higher hypervolume average than the other algorithms. In the case of *sq*1, the IBEA without modifications obtained a higher value compared with the others. It is also possible to see, comparing only

TABLE 8.4: Comparison with the previous works

Inst	M_IBEA	M_NSGA-II	EDA [215]	GA [236]	MMA [140]	ACO [222]	NewACO [223]	PERM [116]
sq1	**−9**	**−9**	**−9**	**−9**	**−9**	**−9**	**−9**	**−9**
sq2	**−9**	**−9**	**−9**	**−9**	**−9**	**−9**	**−9**	**−9**
sq3	**−8**	**−8**	**−8**	**−8**	**−8**	**−8**	**−8**	**−8**
sq4	−13	−13	**−14**	**−14**	**−14**	**−14**	**−14**	**−14**
sq5	**−23**	−22	**−23**	−22	−22	**−23**	**−23**	**−23**
sq6	**−21**	**−21**	**−21**	**−21**		**−21**	**−21**	**−21**
sq7	−35	−34	−35	−34		−34	**−36**	**−36**
sq8	**−42**	−39	**−42**	−37		−32	**−42**	−38

the NSGA-II variants, that the modified version M_NSGA-II obtained better results, except for *sq*1. In general, the MOEAs with backtrack and pool of operators presented an improvement in relation to the traditional MOEAs. In Table 8.3, the cells from M_IBEA that are marked with gray presented statistical differences according to the Kruskal–Wallis test [173] between M_IBEA and all other algorithms (NSGA-II, M_NSGA-II, and IBEA).

8.5.2 Comparison with Previous Single-objective Approaches

This section presents the comparison of the results obtained by the MOEAs with other approaches from the previous works described in Section 8.2.1, and is only concerned with the first objective (Energy of given conformation), since the other works only addressed the single-objective problem. Table 8.4 presents the best results, in terms of energy, found by the modified MOEAs and also the best results obtained by the previous works.

For the first 3 sequences *sq*1, *sq*2, and *sq*3 the modified MOEAs (M_NSGA-II and M_IBEA) obtained the same minimal values of the energy that the previous approaches considered in the comparison. In the case of *sq*4, IBEA, NSGA-II, and M_NSGA-II obtained a value of −13 while M_IBEA and all other optimizers obtained the optimum value of −14. For sequence *sq*5, four single-objective algorithms and M_IBEA have achieved the optimum value −23. However M_NSGA-II and the other algorithms obtained a lesser value of −22. In the case of sequence *sq*6 all algorithms obtained the optimum value of −21. For sequence *sq*7 the M_IBEA obtained −35 as the EDA [215] did. However, the best value found for *sq*7, −36, was obtained by NewACO [223] and PERM [116]. For the last sequence *sq*8 the M_IBEA obtained the optimum value of −42 which is the same obtained by EDA [215] and NewACO [223]. All other approaches obtained lesser values for sequence *sq*8. In general the results obtained by the MOEAs were as good as the best found by other approaches for 6 of the 8 sequences considered.

8.6 Final Remarks

MOEAs are evolutionary algorithms to address the challenge of optimization of multiple objectives at the same time. They have been showing good results in many areas of science. In this chapter two well-known MOEAs were applied in order to address the PSP problem using the HP-2D model. Two multi-objective approaches were presented: the first approach utilizes the standard versions of IBEA and NSGA-II algorithms; the second approach consists of modifying IBEA and NSGA-II, adding backtrack initialization and a pool of operators, in order to enhance the results. Given the experimental results it became clear that the modified versions of the algorithms were able to explore better the search space than the plain approach with standard genetic operators.

Also, it was possible to verify that both multi-objective approaches NSGA-II and M_NSGA-II did not presented satisfactory results when compared with the results achieved by the IBEA and M_IBEA in terms of hypervolume and also when compared with previous single-objective approaches.

The M_IBEA was the algorithm that presented the best results in comparison with NSGA-II, M_NSGA-II, IBEA, and with the previous single-objective studies. This means that a multi-objective formulation alone is not sufficient for achieving good results in terms of energy. Only with the backtrack and the pool of operators the M_IBEA was able to reach acceptable/competitive results. The multi-objective formulation combined with the backtrack initialization, the pool of operators, and the sophisticated mechanism to explore the multi-objective search space of M_IBEA presented promising results. It is arguably that the M_IBEA was able to escape from local optima in almost all cases, except for *sq*7, because the capacity of the multi-objective formulation combined with the pool of operators to generate diversity among the population. Also it is worthwhile to mention that the parameters for the algorithms were not tuned and there is a chance of getting better results if tuning is done for M_IBEA.

The results obtained by the M_IBEA opens a range of possibilities of exploring further multi-objective formulations to the PSP problem within HP-2D model or even for HP-3D. The findings from this study motivates further approaches using multi-objective designs and the addition of a pool of operators in order to enhance the ability of escaping local optima. In the case of multi-objective formulations it is possible to mention that the design of novel approaches, such as using other metrics to measure the compactness of the proteins conformation or other methods that might consider different characteristics, could improve the ability of MOEAs to reach better results.

Future works include to explore alternative selection methods to select the operators from the pool operators and also the addition of more operators. It is believed that the pool of operators is the most responsible for the improvement in the exploration of the MOEAs. Therefore a more intelligent selection mechanism, which considers the history of the operators' applications, could improve even more the performance of the MOEAs. The extension of this work to the HP-3D is also planned. The HP-3D model is more complex than the HP-2D, and it is possible that the multi-objective approach, presented in this chapter, could show a better potential. Finally, the application of hyper-heuristics is also planned for generation of specialized optimization algorithms for the PSP problem.

Chapter 9

Multi-objective IP Assignment for Efficient NoC-based System Design

Maamar Bougherara

Ecole Nationale Supérieure d'Informatique, Algiers, Algeria

Rym Rahmoun

Ecole Normale Supérieure de Kouba, Algiers, Algeria

Amel Sadok

Ecole Normale Supérieure de Kouba, Algiers, Algeria

Nadia Nedjah

State University of Rio de Janeiro University, Rio de Janeiro, Brazil

Mouloud Koudil

Ecole Nationale Supérieure d'Informatique, Algiers, Algeria

Luiza de Macedo Mourelle

State University of Rio de Janeiro University, Rio de Janeiro, Brazil

9.1 Introduction

As the integration of semiconductor increase more complex cores for a system on chip are launched a complex System-on-Chip (SoC) is formed by interconnected heterogeneous component, the interconnection is formed by NoC. A NoC is similar as a general network but with limited resources, power, and area. Each component of the NoC is an intellectual property (IP) block that can have a general or special purpose such as a processor, memory, and DSP [117]. A NoC is designed to run a specific application. This kind of application is generally limited by the number of the tasks that are implemented onto an IP block. An IP Block can implement more than a single task of the application. It can execute many tasks as a general-purpose processor does. In contrast, for instance, a multiplier IP block for floating point numbers can only multiply floating point numbers. The number of IP blocks increase rapidly. So, it is necessary to choose the adequate IP block beforehand as to yield an efficient NoC-based design for any application. The choice becomes harder when the number of task increases. It is also necessary to map these blocks onto the NoC available infrastructure, which consists of a set of cores communicating using switches.

Different optimization criteria can be pursued depending on information details about the application and IP blocks. The application is a graph of tasks called Task Graph (TG). The features of an IP block can be determined from its companion library. The task assigned and IP block mapping are key elements for an efficient NoC-based design [192]. These two problems can be solved using multi-objective optimization. Some of the objectives are conflicting because of their nature. IP assignment and IP mapping are classified as NP-hard problems [97]. The deterministic techniques are not viable to solve such problems. So it is mandatory to use Multi-Objective Evolutionary Algorithms (MOEAs) with specific objective functions. For this purpose, one needs to select the best minimal set of objectives to be optimized. The wrong set of optimized objectives can lead to average instead of the best results.

In this paper, we propose a multi-objective evolutionary based decision to help NoC designers to select the most suited blocks IP used during the assignment step. To this aim, we use the structure representation of TG in [57] and IP repository data from the Embedded Systems Synthesis benchmarks Suite (E3S) [77] as an IP library for the proposed tool. We use the multi-objective particle swarm optimization algorithm, which was modified to suit the specificities of the assignment problem and also to guarantee the NoC designer's constraints [210].

The rest of the paper is organized into eight sections. First, in Section 9.2, we present some related works, in which the assignment stage is viewed as an NP-complete problem. Subsequently, in Section 9.4, we present the model used for the application structure and IP repository. After that, in Section 9.5, we formally present the assignment problem. Then, in Section 9.6, we present the multi-objective particle swarm optimization used in this work. After that, in Section 9.7, we define the objective functions to be optimized by MOPSO. There follows, in Section 9.8, some experimental results and their comparison with some existing work. Last but not least, in Section 9.9, we draw some conclusions and outline new directions for future work.

9.2 Related Work

The problems of allocating IP blocks to application tasks and mapping those blocks into a NoC have been addressed in some previous works. Some of these works treat the assignment and mapping as one single NP-hard problem while others consider them as two distinct NP-hard problems that can be solved separately. Also, some of these works did not take into account the multi-objective nature of the problems and, thus, adopted one single objective. So, before we highlight the works that use IP assignment as one NP-hard problem, we first briefly describe some of the works that are based on an optimization process with one single objective.

Hu and Marculescu [117] proposed a branch and bound algorithm, which automatically maps IPs/cores into a mesh-based NoC architecture minimizing the total amount of consumed power by minimizing the total communication among the used cores. The specified constraints through bandwidth reservation were defined to control communication limits.

Lei and Kumar [147] proposed a two step genetic algorithm for mapping the application into a mesh-based NoC architecture that minimizes the execution time. In the first step, it was assumed that all communication delays are the same and the selection of IP blocks was based on the computation delay imposed by the IPs only. In the second step, they used real communication delays to find an optimal binding of each task in the TG to specific cores of the NoC.

Murali and De Micheli [182] addressed the problem under the bandwidth constraint with the aim of minimizing communication delay by exploiting the possibility of splitting traffic among various paths. Splitting the traffic increases the size of the routing component at each node. However, the size was not one of the concerns in this work.

In the following, we describe some works wherein the assignment and mapping problems are handled via multi-objective optimization.

Zhou et al. [257] suggested a multi-objective exploration approach, treating the mapping problem as two conflicting objective optimization problems that attempts to minimize the average number of hops and achieve a thermal balance. The number of hops is incremented every time data cross a switch before reaching its target. The NSGA multi-objective evolutionary algorithm was used in this case. The authors formulated a thermal model to avoid hot spots, which are areas with high computing activity.

Jena and Sharma [127] addressed the problem of topological mapping of IPs/cores into a mesh-based NoC in two systematic steps using NSGA-II. The main concern was to obtain a solution that minimizes the energy consumption due to both computational and communicational activities and also minimizes the link bandwidth requirements under some prescribed performance constraints.

Da Silva et al. [57] devised an efficient IP assignment multi-objective optimization for NoC platforms. They did so using both NSGA-II and microGA. The work's main concern was the optimization of hardware area, power consumption, and execution time.

Radu [205] used three methods to assign tasks into IP block: (i) random assignment, wherein many tasks can be assigned to the same IP; (ii) direct assignment, wherein a task is assigned to the first IP that can run it; (iii) minimum assignment, wherein the selection is driven by the minimum run time, and thus it always assigns a task to the IP that runs the fastest. After that, Simulated Annealing is used to perform mapping aiming at minimizing the energy consumption.

In [206], SPEA-II and NSGA-II are used with some different crossover and mutation operators for application mapping. The work aimed at optimizing two objectives, which are energy consumption and thermal balance.

9.3 NoC Internal Structure

A NoC platform consists of architecture and design methodology, which scales from a few dozens to several hundreds or even thousands of resources [147]. As mentioned before, a resource may be a processor core, DSP core, an FPGA block, a dedicated hardware block, mixed signal block, memory block of any kind such as RAM, ROM, or CAM, or even a combination of these blocks.

A NoC consists of a set of *resources* (R) and *switches* (S). Resources and switches are connected by *links*. The pair (R, S) forms a *tile*. The simplest way to connect the available resources and switches is arranging them as a mesh so these are able to communicate with each other by sending messages via an available path. A switch is able to buffer and route messages between resource. Each switch is connected to up to four other neighboring switches through input and output channels. While a channel is sending data another channel can buffer incoming data. Figure 9.1 shows the architecture of a mesh-based NoC where each resource contains one or more IP blocks (RNI for resource network interface, D for DSP, M for memory, C for cache, P for processor, FP for floating-point unit, and Re for reconfigurable block). Besides

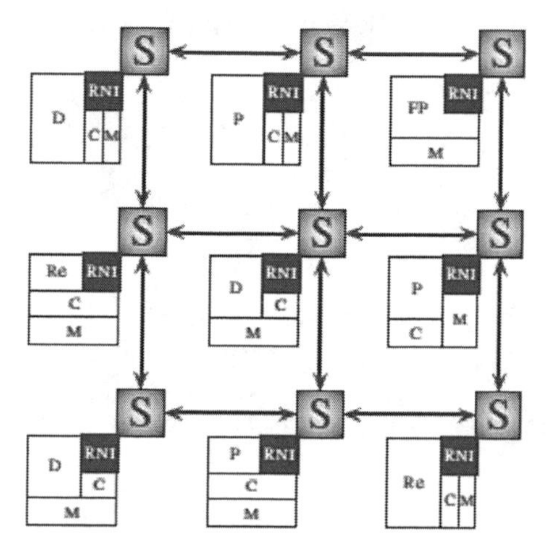

FIGURE 9.1: Mesh-based NoC with nine resources.

the mesh topology, there are more complex topologies like *torus*, *hypercube*, 3-*stage clos*, and *butterfly*. Note that every resource in the NoC must have a unique identifier and is connected to the network via a switch. It communicates with the switch through the available RNI. Thus, any set of IP blocks can be plugged into the network if its footprint fits into an available resource and if this resource is equipped with an adequate RNI.

9.4 Application and IP Repository Models

In order to formulate the IP assignment problem, it is necessary to define the application internal model that will be used. An application can be represented by a set of tasks that can be executed sequentially or in parallel. It is represented by a directed acyclic graph, called Task Graph (TG) [57].

Definition 1 [57]: A task graph $G = (T, D)$ is a directed graph, wherein a node represents a task $t_i \in T$ and a directed arc $d_{ij} \in D$ between tasks t_i, and t_j represents the data dependency between these tasks. The arc label $v(d_{ij})$ characterizes the volume of bits exchanged in communication between tasks t_i and t_j.

The IP assignment determines the association between each task of the application and the IP block that would execute that task. The result of this step is another graph of IP representing the IPs used to implement the application. This graph is called an Application Characterization Graph (ACG).

Definition 2 [57]: An application characterization graph $G = G(C, A)$ is a directed acyclic graph, wherein each vertex $c_i \in C$ represents a selected IP block and each directed arc $a_{ij} \in A$ characterizes the communication process between c_i and c_j.

In order to help NoC designers and standardize the proposed solution, we structured the used application repository, which is the E3S benchmark suite [51] using XML. XML schema provides a neat and well-accepted model for the task graph and IP repository.

9.4.1 Task Graph Representation

A TG is divided into three major elements: The task graph element is the TG itself, which contains tasks and edges. A task element includes a task element for each task of the TG while an edge element includes an edge element for each arc in the TG. Each task has two main attributes: a unique identifier (*code*) and a task type (*type*), chosen among the 46 different types of tasks included in the E3S library [76]. Each edge has four main attributes: a unique identifier (*id*), an identifier of its source node (*src*), another of its target node (*tgt*), and an attribute representing the communication cost imposed (*cost*). Figure 9.2 shows a simple TG of E3S and in Figure 9.3 its corresponding XML representation.

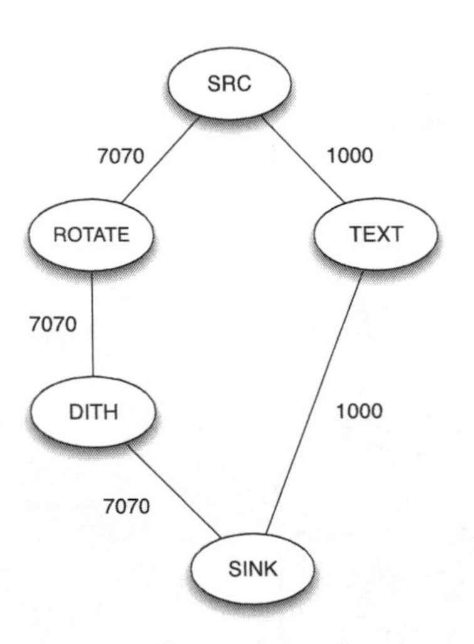

FIGURE 9.2: Example of task graph.

```
<?xml version="1.0" encoding="ISO-8859-1"?>
<task_graph>
  <tasks>
    <task code ="0" name="src"     type="45" />
    <task code ="1" name="text"    type="44" />
    <task code ="2" name="sink"    type="45" />
    <task code ="3" name="rotate" type="43"/>
    <task code ="4" name="dith"    type="42" />
  </tasks>
  <EDGES>
    <edge name="a0_0" from="0" to="1" cost="1000"/>
    <edge name="a0_1" from="0" to="3" cost="7070"/>
    <edge name="a0_2" from="3" to="4" cost="7070"/>
    <edge name="a0_3" from="4" to="2" cost="7070"/>
    <edge name="a0_4" from="1" to="2" cost="1000"/>
  </edges>
</task_graph>
```

FIGURE 9.3: XML code for the example of TG.

9.4.2 Repository Representation

The IP repository is divided into two major elements: the repository and the IP elements. The repository is the IP repository itself. It is noteworthy to point out that the repository contains different non-general purpose embedded processors and each processor implements up to 46 different types of operations. Not all 46 different types of operations are available in all processors. Each type of operation available to be run in each processor is represented by an IP element. Each IP is identified by its unique identifier. It also includes other attributes such as taskType, taskName, taskPower, taskTime, processorID, processorName, processorWidth, processorHeight, processorClock, processorIdle, Area, Power, and cost. The common element in TG and IP repository representations is the type attribute. Therefore, this element will be used to link an IP to a node. Figure 9.4 shows a simplified XML structure representing an IP repository. The repository contains IPs for digital signal processing, matrix operations, text processing, and image manipulation.

9.5 The IP Assignment Problem

IP Assignment is the first step before mapping the application onto NoC [185]. The objective of IP Assignment is to select, from an IP library (IP repository), a set of IPs that exploit re-usability and optimize the implementation of a given application in terms of time, power, and area requirements. During this step, no information about physical location of IPs onto NoC is

```
<?xml version="1.0" encoding="ISO-8859-1"?>
<repository>
  <ips>
    <ip procName="AMD_ElanSC520-133_MHz--square"
     price="33.0" taskTime="9e-06" taskPower="1.6"
     area="9.61E-6" taskName="Angle2Time Conversion"
     type="0" procID="0" id="0"
   />
   <ip procName="AMD_ElanSC520-133_MHz--square"
    price="33.0" taskTime="2.3e-05" taskPower="1.6"
    area="9.61E-6" taskName="Basic floating point"
    type="1" procID="0" id="1"
   />
   <ip procName="AMD_ElanSC520-133_MHz--square"
    price="33.0" taskTime="0.00049"taskPower="1.6"
    area="9.61E-6" taskName="Bit Manipulation"
    type="2" procID="0"id="2"
   />
    . . .
  </ips>
</repository>
```

FIGURE 9.4: XML code for an example of an IP repository.

given. The optimization process must be done based on TG and IP features only. The result of this step is a set of IPs that should maximize the NoC performance, i.e. minimize power consumption, hardware resources as well as the total time execution of the application. Recall that the result of this step is produced in the form of an ACG for the application's task graph, wherein each task has an IP associated with it. The dynamics of this assignment step is illustrated in Figure 9.5.

Note that the number of possible assignments is defined as in Equation 9.1:

$$\#A = n_1 \times n_2 \times \cdots \times n_{m-1} \times n_m, \tag{9.1}$$

wherein m represents the number of tasks used in the application and n_i the number of IPs that can be assigned to task i.

9.6 Assignment with MOPSO Algorithm

The successful application of PSO in many single objective optimization problems reflects the effectiveness and robustness of PSO. The PSO algorithm for a single objective as proposed in [253] is a population-based search algorithm that simulates the social behavior of birds within a flock: the particle simulates the bird's movement while the particle swarm mimics that of the

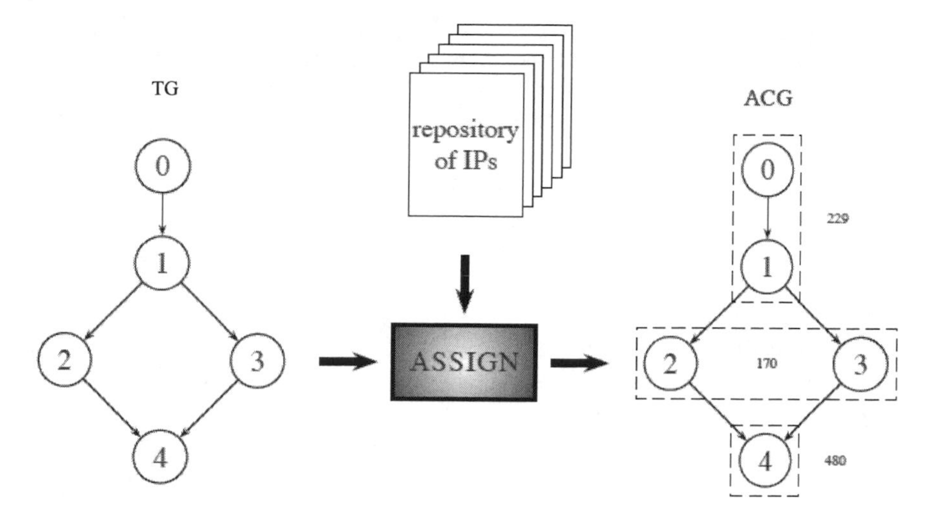

FIGURE 9.5: Dynamics of the assignment process.

flock. In PSO, each particle position represents a potential solution of the problem.

The particle position x_i is a vector of as many coordinates as the dimensions of the search space. It is updated every iteration t using the velocity information as described in Equation 9.2:

$$x_i(t+1) = x_i(t) + v_i(t+1). \tag{9.2}$$

The velocity drives the optimization process to the next position. It is computed as described in Eq. 9.3:

$$\begin{aligned} v_i(t+1) = \quad & w.v_i(t) + \\ & c_1.r_1(t+1)(Pbest(t) - x_i(t)) + \\ & c_2 r_2(t+1)(Gbest(t) - x_i(t)). \end{aligned} \tag{9.3}$$

Three terms are needed to compute the velocity: (i) the inertia, which is based on the previous velocity $w.v_i(t)$. It is used to prevent the particle from suffering a drastic change in direction, where w is the inertia coefficient; (ii) the cognitive term $c_1.r_1(t)(Pbest(t) - x_i(t))$ quantifies the performance of particle i with respect to its previous performance. This component is also known as the "nostalgia" of the particle [186], where c_1 is the cognitive coefficient, $Pbest(t)$ is the best position founded by particle i so far; (iii) the social component, $c_2.r_2(t)(Gbest(t) - x_i(t))$ quantifies the performance of particle (i) with respect to the performance of the swarm of particles. The effect of this term is to attract the particle to the best position found by the particles swarm $Gbest(t)$. The weight c_2 is the social coefficient. The cognitive coefficient, c_1, and the social coefficient, c_2, yield a better performance when these are balanced [186].

Note that $r_1(t)$ and $r_2(t)$ are random numbers selected in the range of [0, 1] at each iteration t. It is used to characterize the stochastic nature of the cognitive and social contributions.

The PSO algorithm cannot be immediately applied to the multi-objective optimization problem. In single-objective optimization, the best solution that each particle uses to update its position is completely determined. However, in multi-objective optimization problems, each particle might have a set of different best solutions, also known as leaders, from which just one has be selected in order to update the particle's position. Such set of leaders is usually stored in a different memory, which is distinct of that of the swarm. This memory is identified as the external archive. This is a repository in which the non-dominated solutions found so far are stored. The solutions contained in the external archive are used as leaders when the positions of the particles of the swarm have to be updated. Furthermore, the contents of the external archive are also usually reported as the final results of the algorithm.

In this paper, we improved the PSO method to a multi-objective approach. The important part in multi-objective particle swarm optimization (MOPSO) is to determine the best global particle for each particle i of the swarm. To facilitate the selection of best and local best, two archives are maintained in addition to the current state of the swarm. One archive stores the global best individuals found so far by the search process, and the other bookkeeps the set of local best positions for each member of the swarm. In our approach, the selection is random. The main algorithm of our MOPSO is shown in Algorithm 10.

Algorithm 10 The main steps of modified MOPSO

```
Initialize the particle of the swarm
Evaluate all the particles of the swarm
Initialize best solutions in archive of leaders
Initialize the archive p_Leader
Initialize the velocity for each particle
iteration ← 0
while iteration < max_iteration do
  for each particle do
    Select a p_leader
    Update Position using Eqs 9.2 and 9.3
    Evaluate particle fitness
    Update p_Leader
  end for
  Update the leaders' archive
  iteration ← iteration + 1
end while
Return result from the archive of leaders
```

The PSO is usually used for the optimization of continuous function. However, in this work we use a discrete version of the multi-objective PSO.

The search space in the assignment problem is defined by the number of tasks m in the application and that of the number n_t of processors that can execute task t. Thus, the number of dimensions of the search space coincides with m, i.e. there is a dimension for each task in the application's TG. Moreover, the magnitude of the search space in each dimension is defined by the n_t, where t is the task associated with the considered dimension. For instance, if the application's TG is composed by 3 tasks, say t_1, t_2, and t_3 that can be executed by 4, 6, and 3 different processes, respectively, then the search space would have 3 distinct dimensions with magnitude of [0, 4], [0, 6], and [0,3], respectively. Furthermore, a confinement strategy is used to keep the particle position within the range of solution space.

9.7 Objective Functions

Different objectives may be considered in the IP assignment problem. The objectives can be concurrent or collaborative [57]. Concurrent objectives should not be grouped, and considered separately during optimization processes. Thus, the process can be treated as a multi-objective optimization.

The best solution for multi-objective optimization is the solution with the adequate trade-off between all concurrent objectives. In this paper, we adopt a multi-objective optimization minimizing three objectives: area, power consumption, and execution time. These objectives must be computed. Thus, we need to provide an executable definition for each of them.

9.7.1 Area

In order to calculate the area required by a given assignment solution, it is necessary to know the area of each of the selected processors or the processing element (pe). As a processor can execute more than one task, we must visit each task of the TG and associate it with the selected processor.

The area is computed summing up all the processor's area. However, when the same processor is associated to several tasks of the TG, the area of that processor is added only once. Eq. 9.4 defines how to compute the required area to use for a given assignment A:

$$Area(A) = \sum_{pe \in Proc(ACG_A)} area_{pe}, \tag{9.4}$$

wherein $Proc()$ yields the set of processors used in ACG for the evaluated assignment A and $area_{pe}$ is the area for processor pe in ACG [76].

9.7.2 Power Consumption

In order to calculate the power consumption, the power required by all the tasks are added up, as defined in Eq. 9.5:

$$Power(A) = \sum_{t \in ACG_A; pe \in Proc(ACG_A)} power_i^{pe}, \tag{9.5}$$

wherein $power_i$ is the required power consumption to execute task i on a specified processor pe [57].

9.7.3 Execution Time

In order to compute the execution time imposed, it is necessary to visit all the tasks of the application's TG and schedule them into their own cycle. Thus, a scheduling algorithm should be applied.

This paper presents two kinds of scheduling algorithms [105]: one based on the As-Soon-As-Possible (ASAP) scheduling strategy and the other on the As-Late-As-Possible (ALAP) scheduling strategy. The first algorithm schedules tasks in the earliest possible control step. So, a task is scheduled if and only if all its predecessors are scheduled in earlier control steps, as described in Algorithm 11. The second algorithm schedules tasks in the latest possible control step. So, a task is scheduled if and only if all its successors are scheduled in later control steps. During each execution cycle, whenever there are several tasks to be executed by the same processor, the time must be accumulated. On the other hand, whenever there are several processors operating in the same control step, the longest processor's cycle is added up, as described in Algorithm 12, wherein s_t represents the control step in which task t is currently scheduled. Eq. 9.6 defines the computation of the execution time:

$$Time(A) = \sum_{s \in Steps(TG)} \max_{pe \in s; t \in s} (time_{pe}), \tag{9.6}$$

wherein $time_t^{pe}$ returns the time needed to execute task t by processor pe.

9.8 Results

The E3S benchmarks suite was used to evaluate the proposed algorithm. The suite contains the characteristics of 17 embedded processors. Each processor is characterized by area, the execution time, and power consumption of 46 different tasks. The suite contains also some applications that are usually executed by embedded systems, such as telecommunication, auto industry, and network. The applications are given by their task graph.

Algorithm 11 ASAP scheduling

$T \leftarrow \emptyset$
$s \leftarrow 1$
for $\forall t \in TG|Pred(t) =$ **do**
  Schedule t in step s
  $T \leftarrow T \cup \{t\}$
end for
while $\forall t \in TG|t \notin T \& Pred(t) \in T$ **do**
  $s \leftarrow s + 1$
  Schedule t in step s
  $T \leftarrow T \cup \{t\}$
end while

Algorithm 12 ALAP scheduling

$T \leftarrow \emptyset$
$s \leftarrow 1$
for $\forall t \in TG|Succ(t) =$ **do**
  Schedule t in step s
  $T \leftarrow T \cup \{t\}$
end for
while $\forall t \in TG|t \notin T \& Succ(t) \in T$ **do**
  $s \leftarrow s + 1$
  Schedule t in step s
  $T \leftarrow T \cup \{t\}$
end while
for $\forall t \in TG$ **do**
  $s_t \leftarrow s - s_t + 1$
end for

The parameters used during the simulation are $c_1 = c_2 = 1$ or $c_1 = c_2 = 2$; w was set to either 0.6 or 0.7 and the swarm size and number of iterations of 200, 500, 2000, and 5000 were tested. After several tentative simulations, we selected the best parameter setting as shown in Table 9.1.

TABLE 9.1: Parameter setting

Case	$c_1 = c_2$	w	Swarm size	#Iteration
1	1	0.6	5000	5000
2	2	0.6	5000	5000
3	1	0.7	5000	5000
4	2	0.7	5000	5000

Among the 16 common applications executed in embedded environments, as provided by E3S, we chose five non-linear graphs, which we think of as representative applications. Table 9.2 shows the characteristics of the benchmark applications.

TABLE 9.2: Details of the benchmark applications

Application	ID	m	n	#A
auto-indust-tg2	1	9	9	606,076,928
consumer-tg1	2	7	5	176,868
Office-tg0	3	5	5	210,681
Telecom-tg1	4	6	6	9,516,192
Telecom-tg2	5	6	6	9,516,192

We provide different results for these five applications: when using the ASAP and ALAP scheduling algorithm. Table 9.3 shows the characteristics of the found assignments for the ASAP scheduling strategy while Table 9.4 shows the same characteristics of the found assignments for the ALAP scheduling strategy. Note that the power figures are watts, time is 10^{-3} seconds, and area is 10^{-6} squared meter.

TABLE 9.3: IP assignment using ASAP scheduling

App ID	Best assignment			Average assignment		
	power	time	area	power	time	area
1	7.15	0.8325	1	8.208	0.8332	7.3
2	0.375	14.52	2.44	3.321	39.148	7.591
3	0.375	4.92	1	8.461	303.0969	25.24
4	0.45	0.2097	1	3.937	1.958	54.9
5	0.45	0.213	1	3.955	1.994	10.91

TABLE 9.4: IP assignment using ALAP scheduling

App ID	Best assignment			Average assignment		
	power	time	area	power	time	area
1	7.15	0.832	1	8,208	0.8332	7.38
2	0.375	14.52	2.44	4.17	28.595	8.40
3	0.375	4,22	1	4.72	401.60	14.49
4	0.45	0,213	1	4.279	1.778	9.83
5	0.45	0.213	1	4.325	1.81	9.83

It is clear from these results that the ASAP strategy allowed for better assignment in most of these simulation cases. The results show that ASAP scheduling is more effective than ALAP but not always. So, it would be better to choose the scheduling strategy automatically. The best results are summarized in Table 9.5.

TABLE 9.5: IP assignment for the application in E3S using MOPSO and scheduling strategy used

App	Best assignment				Average assignment			
	power	time	area	S	power	time	area	S
1	7.15	0.832	1	1	8.208	0.832	7.3	1
2	0.375	14.52	2.44	1	3.321	39.148	7.591	1
3	0.375	4.22	1	2	4.72	401.60	14.49	2
4	0.45	0.209	1	1	4.279	1.778	9.83	2
5	0.45	0.213	1	1	4.325	1.81	9.83	2

TABLE 9.6: Samples of non-dominated optimal assignment, found by MOPSO

App	Non-dominated assignment	P	T	A
1	[462, 375, 380, 379, 376, 370, 383, 384, 462]	x		
	[462, 375, 380, 379, 6, 370, 383, 384, 462]		x	
	[386, 375, 380, 379, 376, 370, 383, 384, 386]			x
2	[462, 455, 462, 457, 462]	x		
	[369, 240, 369, 242, 369]		x	
	[462, 455, 462, 457, 462]			x
3	[462, 461, 462, 460, 459]	x		
	[462, 96, 462, 245, 244]		x	
	[462, 461, 462, 367, 366]			x
4	[462, 438, 441, 447, 444, 462]	x		
	[462, 463, 441, 472, 469, 462]		x	
	[201, 185, 188, 194, 191, 201]			x
5	[462, 438, 441, 447, 444, 462]	x		
	[462, 463, 466, 472, 469, 462]		x	
	[479, 463, 466, 472, 469, 479]			x

Table 9.6 shows some of the non-dominated assignments, found by MOSPO, for the selected applications. The given lists show the number of allocated IPs as defined in E3S [184], whereby the position in the list indicates the task number in the application's corresponding task graph. The occurrence of the symbol "x" in the last columns of this table indicates that the associated assignment reached the minimum for the highlighted objective (P for power consumption, A for area, and T for time).

The Pareto fronts of the IP assignment as obtained for the considered applications are shown in Figure 9.6. Note that for each application, the fronts evolved for the scheduling strategies ASAP and ALAP are shown. Also, it is noteworthy to point out that there are some solutions that are common to both fronts.

In order to consolidate the proposed approach, we compare the obtained results to those reported in [57], where the NSGA-II and MicroGA were exploited to yield assignments. It is noteworthy to emphasize that the same objective function is used in the compared works. Table 9.7 shows the charac-

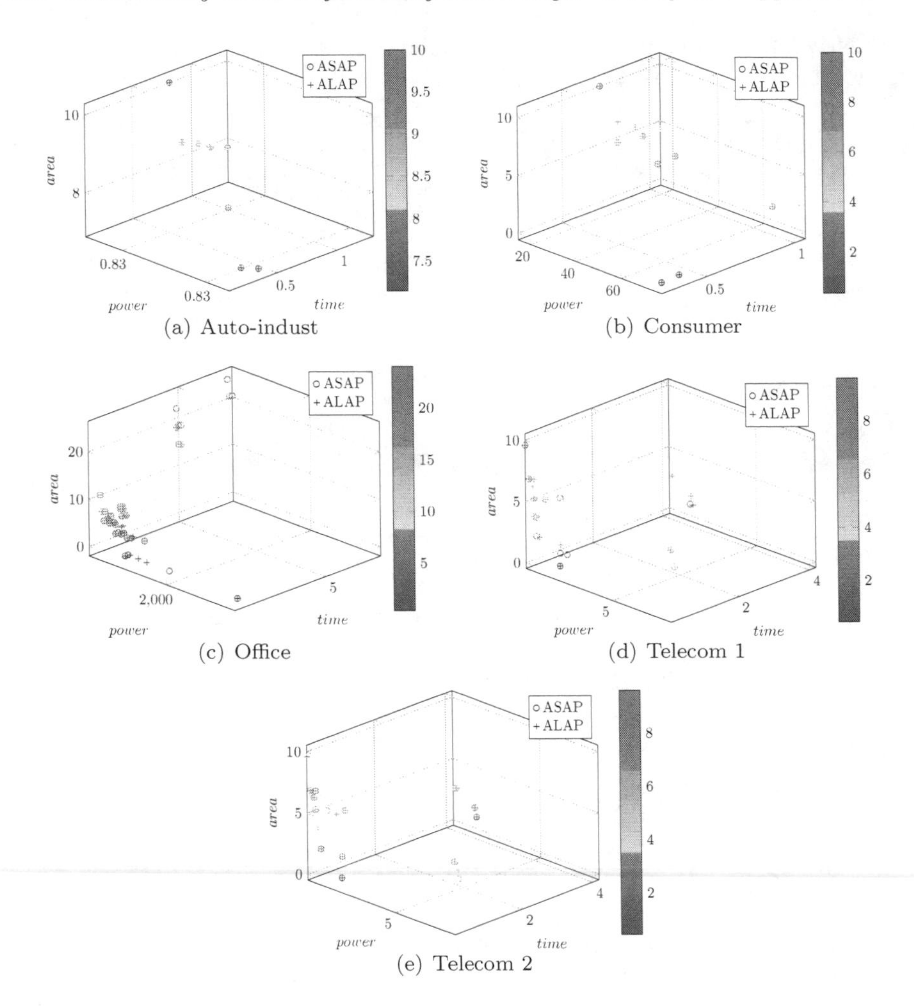

FIGURE 9.6: Illustration of the Pareto fronts obtained for the exploited applications: Telecom 2

teristics of the best assignment using NSGA-II, and Table 9.8 show the results as obtained by MicroGA.

When we compare the obtained results with those reported in [57], as shown in Tables 9.7 and 9.8, it becomes clear that our approach with the MOPSO algorithm provides better results than NSGA and MicroGA. However, considering the time objective, some benchmarks are better served with NSGA-II and MicroGA. The charts of Figure 9.7 allow a visual comparison of the power consumption for the assignments' yield by the compared strategies, regarding best and average results. Also, the charts of Figure 9.8 should permit a better visual comparison of results regarding the time characteristic

TABLE 9.7: IP assignment for the application in E3S using NSGA-II [57]

App	Best assignment			Average assignment		
	power	time	area	power	time	area
1	7.15	0.845	9.182	8.937	2	11.68
2	0.375	14.5	2.488	3.428	47.2	7.878
3	0.375	4	2	4.376	446	15.76
4	0.45	0.219	2	3.414	1	9.498
5	0.45	0.219	2	3.146	1	10.31

TABLE 9.8: IP assignment for the application in E3S using MicroGA [57]

App	Best assignment			Average assignment		
	power	time	area	power	time	area
1	7.75	0.845	9.182	9.819	1	14.2
2	0.375	14	2.48	3.353	43	7.993
3	0.375	4	2	8.489	322	28.6
4	0.3	0.219	2	3.612	1	1
5	0.45	0.219	2	3.681	1	9.785

considering the best and average quality assignments. Finally, the charts of Figure 9.9 facilitate the comparison of the required hardware area that would be occasioned by the assignments obtained by the compared strategies.

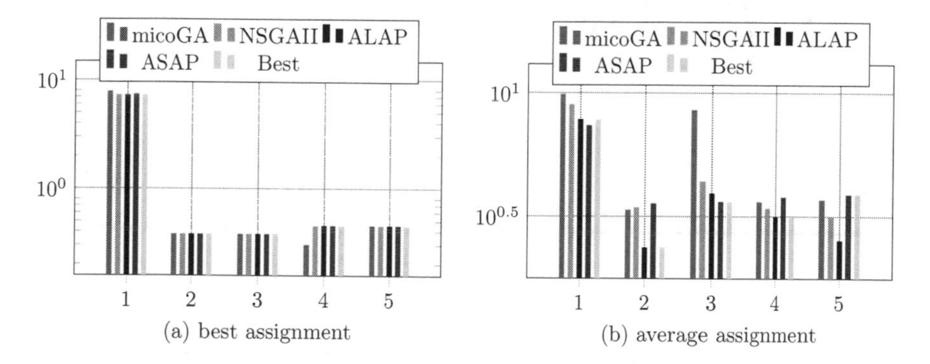

FIGURE 9.7: Comparison of the assignments regarding power requirement.

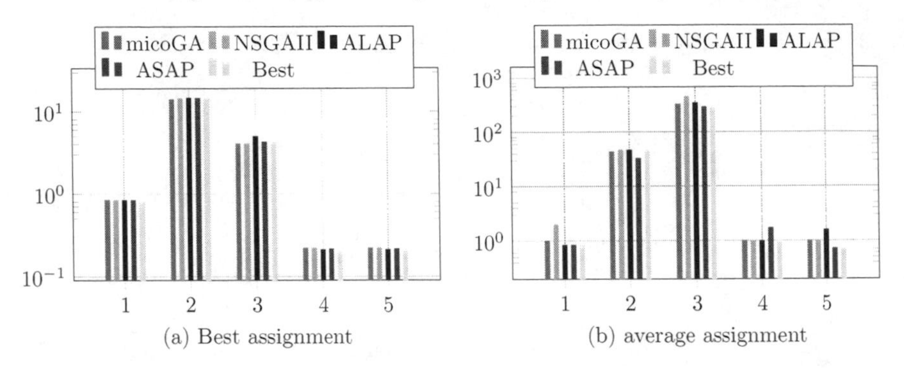

FIGURE 9.8: Comparison of the obtained assignments regarding time requirement

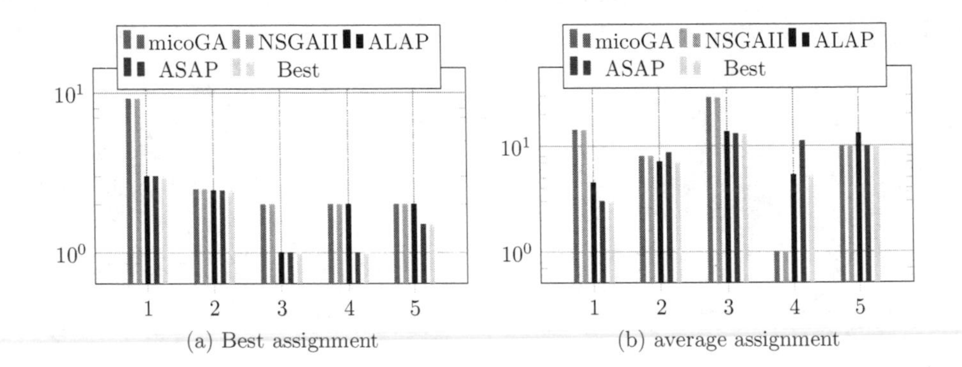

FIGURE 9.9: Comparison of the obtained assignments regarding area requirement

9.9 Conclusions

The IP assignment step is an NP-hard combinatorial problem in NoC design. In this paper we propose an algorithm based on MOPSO to help NoC designers to select a set of adequate IPs from repository IP. We use the structure of TG and ACG presented in [185], and we adopt E3S benchmark as our repository IP. We exploit the same objective used in [57]. Two scheduling algorithms have been tested our tool. The obtained results are better than presented in [57]. For future work, we intend to use the same algorithm to tackle the mapping step.

Bibliography

[1] Mohammad Saniee Abadeh and Jafar Habibi. Computer intrusion detection using an iterative fuzzy rule learning approach. In *FUZZ-IEEE*, pages 1–6. IEEE, 2007.

[2] J. Alcalá-Fdez, A. Fernandez, J. Derrac J. Luengo, L. Sánchez S. García, and F. Herrera. Keel data-mining software tool: Data set repository, integration of algorithms and experimental analysis framework. *Journal of Multiple-Valued Logic and Soft Computing*, 17(2-3):255–287, 2011.

[3] Saleh Alshomrani, Abdullah Bawakid, Seong-O Shim, Alberto Fernández, and Francisco Herrera. A proposal for evolutionary fuzzy systems using feature weighting: Dealing with overlapping in imbalanced datasets. *Knowledge-Based Systems*, 73:1–17, 2015.

[4] Michela Antonelli, Pietro Ducange, and Francesco Marcelloni. An experimental study on evolutionary fuzzy classifiers designed for managing imbalanced datasets. *Neurocomputing*, 146(0):125–136, 2014. Bridging Machine Learning and Evolutionary Computation (BMLEC) Computational Collective Intelligence.

[5] B. Aslam, F. Amjad, and C. Zou. Optimal roadside units placement in urban areas for vehicular networks. In *IEEE Symposium on Computers and Communications*, pages 423–429, July 2012.

[6] K.J. Åström and T. Hägglund. *Advanced PID Control.* ISA-The Instrumentation, Systems, and Automation Society, 2006.

[7] Welding Handbook AWS. *Welding Technology.* 8th edition, 1991.

[8] Meghna Babbar-Sebens and Barbara S. Minsker. Interactive genetic algorithm with mixed initiative interaction for multi-criteria ground water monitoring design. *Applied Soft Computing*, 12(1):182–195, 2012.

[9] T. Bäck, D. Fogel, and Z. Michalewicz, editors. *Handbook of Evolutionary Computation.* Oxford University Press, 1997.

[10] Thomas Bäck. *Evolutionary algorithms in theory and practice: Evolution Strategies, Evolutionary Programming, Genetic Algorithms.* Oxford University Press, New York, 1996.

[11] J. Bader and E. Zitzler. Hype: An algorithm for fast hypervolume-based many-objective optimization. *Evolutionary Computation*, 19(1):45–76, March 2011.

[12] Hamed Badihi, Youmin Zhang, and Henry Hong. Fuzzy gain-scheduled active fault-tolerant control of a wind turbine. *Journal of the Franklin Institute*, 351(7):3677–3706, 2014.

[13] N. Balouchzahi, M. Fathy, and A. Akbari. Optimal road side units placement model based on binary integer programming for efficient traffic information advertisement and discovery in vehicular environment. *IET Intelligent Transport Systems*, 9:851–861, 2015.

[14] David Barber. *Bayesian Reasoning and Machine Learning*. Cambridge University Press, 2012.

[15] Ugo Bastolla, Helge Frauenkron, Erwin Gerstner, Peter Grassberger, and Walter Nadler. Testing a new Monte Carlo algorithm for protein folding. *arXiv preprint cond-mat/9710030*, 1997.

[16] G.E.A.P.A. Batista, R.C. Prati, and M.C. Monard. A study of the behavior of several methods for balancing machine learning training data. *SIGKDD Explorations*, 6(1):20–29, 2004.

[17] Henri Joseph Léon Baudrillart. *Manuel d'économie politique*. Guillaumin et cie, 1872.

[18] Andrea Bazzoli and Andrea G.B. Tettamanzi. A memetic algorithm for protein structure prediction in a 3d-lattice HP model. In *Applications of Evolutionary Computing*, pages 1–10. Springer, 2004.

[19] Slim Bechikh, Lamjed Ben Said, and Khaled Ghédira. Group preference-based evolutionary multi-objective optimization with non-equally important decision makers: Application to the portfolio selection problem. *International Journal of Computer Information Systems and Industrial Management Applications*, 5(278-288):71, 2013.

[20] Khouloud Bedoud, Mahieddine Ali-rachedi, Tahar Bahi, and Rabah Lakel. Adaptive fuzzy gain scheduling of {PI} controller for control of the wind energy conversion systems. *Energy Procedia*, 74:211–225, 2015.

[21] Lamjed Ben Said, Slim Bechikh, and Khaled Ghédira. The r-dominance: A new dominance relation for interactive evolutionary multicriteria decision making. *Evolutionary Computation, IEEE Transactions on*, 14(5):801–818, 2010.

[22] Nicola Beume, Boris Naujoks, and Michael Emmerich. Sms-emoa: Multiobjective selection based on dominated hypervolume. *European Journal of Operational Research*, 181(3):1653–1669, 2007.

[23] Nicola Beume, Boris Naujoks, and Michael Emmerich. Sms-emoa: Multiobjective selection based on dominated hypervolume. *European Journal of Operational Research*, 181(3):1653–1669, 2007.

[24] James C. Bezdek, Robert Ehrlich, and William Full. Fcm: The fuzzy c-means clustering algorithm. *Computers & Geosciences*, 10(2):191–203, 1984.

[25] J.C. Bezdek. *Pattern Recognition with Fuzzy Objective Function Algorithms*. Advanced applications in pattern recognition. Plenum Press, 1981.

[26] L. Bezerra, M. Lopez-Ibanez, and T. Stuetzle. Automatic component-wise design of multi-objective evolutionary algorithms. *Evolutionary Computation, IEEE Transactions on*, PP(99):1–1, 2015.

[27] Leonardo C. T. Bezerra, Manuel López-Ibáñez, and Thomas Stützle. *Evolutionary Multi-Criterion Optimization: 8th International Conference, EMO 2015, Guimarães, Portugal, March 29 –April 1, 2015. Proceedings, Part I*, chapter Comparing Decomposition-Based and Automatically Component-Wise Designed Multi-Objective Evolutionary Algorithms, pages 396–410. Springer International Publishing, Cham, 2015.

[28] E. Biham and A. Shamir. Differential cryptanalysis of des-like cryptosystems. *Journal of Cryptology*, 4(1):3–72, 1991.

[29] Peter A.N. Bosman and Dirk Thierens. The balance between proximity and diversity in multi-objective evolutionary algorithms. *Evolutionary Computation, IEEE Transactions on*, 7(2):174–188, 2003.

[30] Mohamed Ben Brahim, Wassim Drira, and Fethi Filali. Roadside units placement within city-scaled area in vehicular ad-hoc networks. In *3rd International Conference on Connected Vehicles and Expo*, pages 1–7. IEEE, 2014.

[31] C. Bunkhumpornpat, K. Sinapiromsaran, and C. Lursinsap. Safe-level-smote: Safe-level-synthetic minority over-sampling technique for handling the class imbalanced problem. In *Pacific-Asia Conference on Knowledge Discovery and Data Mining(PAKDD09)*, volume 5476 of *Lecture Notes on Computer Science*, pages 475–482. Springer-Verlag, 2009.

[32] C. Edward Hinojosa, Heloisa A. Camargo, and V. Yván J. Túpac. Learning fuzzy classification rules from imbalanced datasets using multi-objective evolutionary algorithm. *Latin American Congress on Computational Intelligence (LA-CCI)*, 1(1), 2015.

[33] C2C-CC. CAR 2 CAR Communication Consortium Manifesto. Technical report, CAR 2 CAR Communication Consortium, 2007. [online], available in `https://www.car-2-car.org/index.php?id=31`.

[34] William D. Callister and David G. Rethwisch. *Materials Science and Engineering: An Introduction*, volume 7. Wiley, New York, 2007.

[35] C. Campolo, A. Molinaro, and R. Scopigno, editors. *Vehicular Ad Hoc Networks - Standards, Solutions, and Research.* Springer, 2015.

[36] Hong Cao, Xiao-Li Li, D.Y.-K. Woon, and See-Kiong Ng. Integrated oversampling for imbalanced time series classification. *Knowledge and Data Engineering, IEEE Transactions on*, 25(12):2809–2822, Dec 2013.

[37] E.H. Cardenas and H.A. Camargo. Multi-objective genetic generation of fuzzy classifiers using the iterative rule learning. In *Fuzzy Systems (FUZZ-IEEE), 2012 IEEE International Conference on*, pages 1–8, June 2012.

[38] Howard B. Cary and Scott C. Helzer. *Modern Welding Technology.* 6th edition, Prentice-Hall, 2004.

[39] E. Cavalcante, A. Aquino, G. Pappa, and A. Loureiro. Roadside unit deployment for information dissemination in a VANET: An evolutionary approach. In *14th Genetic and Evolutionary Computation Conference*, pages 27–34, 2012.

[40] Cetacea. Cetacea Wireless Solutions Company shop. Online at https://shop.cetacea.com/. Retrieved December 2015.

[41] H.S. Chan and E. Bornberg-Bauer. Perspectives on protein evolution from simple exact models. *Applied Bioinformatics*, 1(3):121–144, 2001.

[42] Nitesh V. Chawla, Kevin W. Bowyer, Lawrence O. Hall, and W. Philip Kegelmeyer. Smote: Synthetic minority over-sampling technique. *Journal of Artificial Intelligence Research*, 16:321–357, 2002.

[43] H. Cheng, X. Fei, A. Boukerche, A. Mammeri, and M. Almulla. A geometry-based coverage strategy over urban VANETs. In *10th ACM Symposium on Performance Evaluation of Wireless Ad Hoc, Sensor, & Ubiquitous Networks*, pages 121–128, 2013.

[44] B. Chew. *Hydrogen Control of Basic-coated MMA Welding Electrodes: Part 3: Relationship Between Coating Moisture and Weld Hydrogen.* Central Electricity Generating Board, Marchwood (United Kingdom). Research Div., 1981.

[45] C. Coello, D. Van Veldhuizen, and G. Lamont. *Evolutionary Algorithms for Solving Multi-objective Problems.* Kluwer, New York, 2002.

[46] C. A. Coello and G. B. Lamont. *Applications of Multi-objective Evolutionary Algorithms.* London: World Scientifics, 2004.

[47] C. A. C. Coello, G.L. Lamont, and D.A. van Veldhuizen. *Evolutionary Algorithms for Solving Multi-Objective Problems.* Genetic and Evolutionary Computation. 2nd edition, Springer, Berlin, Heidelberg, 2007.

[48] Carlos Coello Coello, Gary B. Lamont, and David A. Van Veldhuizen. *Evolutionary Algorithms for Solving Multi-objective Problems.* Springer Science & Business Media, 2007.

[49] Yann Collette and Patrick Siarry. *Multi-objective Optimization: Principles and Case Studies.* Springer, 2003.

[50] W.J. Conover. *Practical Nonparametric Statistics.* John Wiley and Sons, New York, 1999.

[51] World Wide Web Consortium(W3C). The world wide web consortium, `http://www.w3.org`, 2008.

[52] O. Cordón. *Genetic Fuzzy Systems: Evolutionary Tuning and Learning of Fuzzy Knowledge Bases.* Advances in fuzzy systems - Applications and theory. World Scientific, 2001.

[53] Oscar Cordon, Fernando A. C. Gomide, Francisco Herrera, Frank Hoffmann, and Luis Magdalena. Ten years of genetic fuzzy systems: Current framework and new trends. *Fuzzy Sets and Systems*, 141(1):5–31, 2004.

[54] Carlos Cotta. Protein structure prediction using evolutionary algorithms hybridized with backtracking. In *Artificial Neural Nets Problem Solving Methods*, pages 321–328. Springer, 2003.

[55] Fábio L. Custódio, Hélio J.C. Barbosa, and Laurent E Dardenne. Investigation of the three-dimensional lattice HP protein folding model using a genetic algorithm. *Genetics and Molecular Biology*, 27(4):611–615, 2004.

[56] Fábio Lima Custódio, Helio J.C. Barbosa, and Laurent Emmanuel Dardenne. A multiple minima genetic algorithm for protein structure prediction. *Applied Soft Computing*, 15:88–99, 2014.

[57] Marcus Vinícius Carvalho da Silva, Nadia Nedjah, and Luiza de Macedo Mourelle. Evolutionary ip assignment for efficient noc-based system design using multi-objective optimization. In *IEEE Congress on Evolutionary Computation*, pages 2257–2264. IEEE, 2009.

[58] Swagatam Das, Sankha Subhra Mullick, and P.N. Suganthan. Recent advances in differential evolution – an updated survey. *Swarm and Evolutionary Computation*, pages –, 2016.

[59] Dayvid Victor Rodrigues de Oliveira. *Análise comparativa de algoritmos de seleção de protótipos em bases desbalanceadas.* Centro de Informática, 2009.

[60] K. Deb. *Multi-Objective Optimization Using Evolutionary Algorithms.* England: Wiley, 2001.

[61] K. Deb. *Multi-Objective Optimization Using Evolutionary Algorithms.* Wiley-Interscience Series in Systems and Optimization. John Wiley & Sons, Chichester, 2001.

[62] K. Deb, A. Pratap, S. Agarwal, and T. Meyarivan. A fast and elitist multi-objective genetic algorithm: NSGA-II. *Trans. Evol. Comp*, 6(2):182–197, April 2002.

[63] K. Deb, A. Pratap, S. Agarwal, and T. Meyarivan. A fast and elitist multi-objective genetic algorithm: NSGA-II. *IEEE Transactions on Evolutionary Computation*, 6(2):182–197, 2002.

[64] K. Deb, A. Pratap, S. Agarwal, and T. Meyarivan. A fast and elitist multi-objective genetic algorithm: NSGA-II. *IEEE J. EVC*, 6(2):182–197, 2002.

[65] Kalyanmoy Deb. *Multi-Objective Optimization Using Evolutionary Algorithms*, volume 16. John Wiley & Sons, 2001.

[66] Kalyanmoy Deb and Ram Bhushan Agrawal. Simulated binary crossover for continuous search space. *Complex Systems*, 1995.

[67] Kalyanmoy Deb and Abhay Kumar. Light beam search based multi-objective optimization using evolutionary algorithms. In *Evolutionary Computation, 2007. CEC 2007. IEEE Congress on*, pages 2125–2132. IEEE, 2007.

[68] Kalyanmoy Deb and Abhishek Kumar. Interactive evolutionary multi-objective optimization and decision-making using reference direction method. In *Proceedings of the 9th Annual Conference on Genetic and Evolutionary Computation*, pages 781–788. ACM, 2007.

[69] Kalyanmoy Deb, Amrit Pratap, Sameer Agarwal, and TAMT Meyarivan. A fast and elitist multi-objective genetic algorithm: Nsga-ii. *Evolutionary Computation, IEEE Transactions on*, 6(2):182–197, 2002.

[70] Kalyanmoy Deb, Ankur Sinha, Pekka J Korhonen, and Jyrki Wallenius. An interactive evolutionary multi-objective optimization method based on progressively approximated value functions. *Evolutionary Computation, IEEE Transactions on*, 14(5):723–739, 2010.

[71] Kalyanmoy Deb, J. Sundar, N. Udaya Bhaskara Rao, and Shamik Chaudhuri. Reference point based multi-objective optimization using evolutionary algorithms. *International Journal of Computational Intelligence Research*, 2(3):273–286, 2006.

[72] Kalyanmoy Deb, Lothar Thiele, Marco Laumanns, and Eckart Zitzler. Scalable test problems for evolutionary multi-objective optimization. Technical report, Computer Engineering and Networks Laboratory (TIK), Swiss Federal Institute of Technology (ETH), 2001.

[73] Kalyanmoy Deb, Lothar Thiele, Marco Laumanns, and Eckart Zitzler. *Scalable Test Problems for Evolutionary Multi-Objective optimization.* Springer, 2005.

[74] Kalyanmoy Deb and Santosh Tiwari. Omni-optimizer: A generic evolutionary algorithm for single and multi-objective optimization. *European Journal of Operational Research*, 185(3):1062–1087, 2008.

[75] Alexandre Cláudio Botazzo Delbem. *Restabelecimento de energia em sistemas de distribuição por algoritmo evolucionário associado a cadeias de grafos.* PhD thesis, Universidade de São Paulo, 2002.

[76] Robert P. Dick. Embedded system synthesis benchmarks suite (e3s). `http://ziyang.eecs.northwestern.edu/dickrp/e3s/`, 2016.

[77] Robert P. Dick, David L. Rhodes, and Wayne Wolf. Tgff: Task graphs for free. In *6th International Workshop on Hardware/Software Co-design*, pages 97–101. IEEE, 1998.

[78] Adrián Domínguez, Joseba Saenz-de Navarrete, Luis De-Marcos, Luis Fernández-Sanz, Carmen Pagés, and José-Javier Martínez-Herráiz. Gamifying learning experiences: Practical implications and outcomes. *Computers & Education*, 63:380–392, 2013.

[79] W. Dong, Z. Qie, Z. Zhou, X. Wu, M. Liu, and W. Zheng. The optimization of mixture ratio of model sand based on simplex particle swarm optimization algorithm. In *Chinese Control and Decision Conference - CCDC'2008*, pages 3868–3872, Yantai, China, 2008.

[80] M. Dorigo and M. Maniezzo. Parallel genetic algorithms: Introduction and overview of current research. In *Parallel Genetic Algorithms*, J. Stender (Ed.). IOS Press, 1993.

[81] Juan J. Durillo and Antonio J. Nebro. jMetal: A java framework for multi-objective optimization. *Advances in Engineering Software*, 42(10):760–771, 2011.

[82] Matthias Ehrgott. *Multicriteria Optimization*, volume 2. Springer, 2005.

[83] Matthias Ehrgott. A discussion of scalarization techniques for multiple objective integer programming. *Ann. Ope. Res.*, 147:343–360, 2006.

[84] Michael Emmerich, Nicola Beume, and Boris Naujoks. An emo algorithm using the hypervolume measure as selection criterion. In *Evolutionary Multi-Criterion Optimization*, pages 62–76. Springer, 2005.

[85] M. Fazzolari, R. Alcala, Y. Nojima, H. Ishibuchi, and F. Herrera. A review of the application of multi-objective evolutionary fuzzy systems: Current status and further directions. *Fuzzy Systems, IEEE Transactions on*, 21(1):45–65, Feb 2013.

[86] FCC. Federal Communications Commission news. [online], available in `http://transition.fcc.gov/Bureaus/Engineering_Technology/News_Releases/1999/nret9006.html`, 1999. Retrieved January 2016.

[87] FCC. Connected Vehicle Research in the United States. [online], available in `http://www.its.dot.gov/connected_vehicle/connected_vehicle_research.htm`, 2015. Retrieved January 2016.

[88] Gang Feng. A survey on analysis and design of model-based fuzzy control systems. *Fuzzy Systems, IEEE Transactions on*, 14(5):676–697, Oct 2006.

[89] A. Fernandez, M.J. del Jesus, and F. Herrera. Sistemas basados en reglas difusas en clasificación: Nuevos retos. In *XIV Congreso Espa nol sobre Tecnologías y Lógica Fuzzy(ESTYLF08)*, pages 493–500, 2008.

[90] Alberto Fernández, María José del Jesus, and Francisco Herrera. On the 2-tuples based genetic tuning performance for fuzzy rule based classification systems in imbalanced data-sets. *Information Sciences*, 180(8):1268–1291, 2010.

[91] Elliackin Messias do Nascimento Figueiredo, Teresa Bernarda Orientadora Ludermir, and Carmelo José Albanez Coorientador Bastos Filho. Algoritmo baseado em enxame de partículas para otimização de problemas com muitos objetivos. *Repositório Institucional da UFPE*, 2013.

[92] Carlos M. Fonseca, Joshua D. Knowles, Lothar Thiele, and Eckart Zitzler. A tutorial on the performance assessment of stochastic multiobjective optimizers. In *Third International Conference on Evolutionary Multi-Criterion Optimization (EMO 2005)*, volume 216, page 240, 2005.

[93] Paulo H.R. Gabriel, Vinícius V. de Melo, and Alexandre C.B. Delbem. Algoritmos evolutivos e modelo hp para predição de estruturas de proteínas. *Revista de Controle e Automação*, 23(1):25–37, 2012.

[94] S.M. Gaffer, M.E. Yahia, and K. Ragab. Genetic fuzzy system for intrusion detection: Analysis of improving of multiclass classification accuracy using kddcup-99 imbalance dataset. In *Hybrid Intelligent Systems (HIS), 2012 12th International Conference on*, pages 318–323, Dec 2012.

[95] V. Ganganwar. An overview of classification algorithms for imbalanced datasets. *International Journal of Emerging Technology and Advanced Engineering*, 2(4):42–47, 2012.

[96] Mostafa Fathi Ganji, Mohammad Saniee Abadeh, Mahdi Hedayati, and N. Bakhtiari. Fuzzy classifcation of imbalanced data sets for medical diagnosis. In *Iranian Conference of Biomedical Engineering*, 2010.

[97] M. R. Garey and D. S. Johnson. *Computers and Intractability; A Guide to the Theory of NP-Completeness*. Freeman and Company, 1979.

[98] Fred Glover. Future paths for integer programming and links to artificial intelligence. *Computers & Operations Research*, 13(5):533–549, 1986.

[99] David E. Goldberg. *Genetic Algorithms in Search, Optimization and Machine Learning*. Addison-Wesley Longman Publishing Co., Inc., Boston, MA, USA, 1st edition, 1989.

[100] Maoguo Gong, Fang Liu, Wei Zhang, Licheng Jiao, and Qingfu Zhang. Interactive moea/d for multi-objective decision making. In *Proceedings of the 13th Annual Conference on Genetic and Evolutionary Computation*, pages 721–728. ACM, 2011.

[101] A. Gonzalez and R. Perez. Improving the genetic algorithm of slave. *Mathware and Soft Computing*, 16(1):59–70, 2009.

[102] Antonio Gonzalez and Raul Perez. Slave: A genetic learning system based on an iterative approach. *IEEE Transactions on Fuzzy Systems*, 7:176–191, 1999.

[103] David Perry Greene and Stephen F. Smith. Competition-based induction of decision models from examples. *Machine Learning*, 13(2-3):229–257, 1993.

[104] Charles Miller Grinstead and James Laurie Snell. *Introduction to Probability*. American Mathematical Soc., 2012.

[105] F. Gruian and K. Kuchcinski. Lenes: Task scheduling for low-energy systems using variable supply voltage processors. In *Proceedings of the Asia and South Pacific Design Automation Conference*, pages 449–455. ACM, 2001.

[106] H. Han, W.Y. Wang, and B.H. Mao. Borderline-smote: a new oversampling method in imbalanced data sets learning. In *2005 International Conference on Intelligent Computing(ICIC05)*, volume 3644 of *Lecture Notes on Computer Science*, pages 878–887. Springer-Verlag, 2005.

[107] H. Hartenstein and K. Laberteaux. *VANET Vehicular Applications and Inter-Networking Technologies*. Intelligent Transport Systems. John Wiley & Sons, Upper Saddle River, NJ, USA, December 2009.

[108] R. L. Haupt and S. E. Haupt. *Practical Genetic Algorithms*. John Wiley, 1998.

[109] Simon O. Haykin. *Neural Networks and Learning Machines*, volume 3. Prentice Hall, 2008.

[110] H. He, Y. Bai, E.A. Garcia, and S. Li. Adasyn: Adaptive synthetic sampling approach for imbalanced learning. In *2008 International Joint Conference on Neural Networks(IJCNN08)*, pages 1322–1328, 2008.

[111] F. Herrera. Genetic fuzzy systems: Status, critical considerations and future directions. *International Journal of Computational Intelligence Research*, 1(1):59–67, 2005.

[112] Francisco Herrera. Genetic fuzzy systems: Taxonomy, current research trends and prospects. *Evolutionary Intelligence*, 1(1):27–46, 2008.

[113] T. Ryan Hoens, Qi Qian, Nitesh V. Chawla, and Zhi-Hua Zhou. Building decision trees for the multi-class imbalance problem. In Pang-Ning Tan, Sanjay Chawla, ChinKuan Ho, and James Bailey, editors, *Advances in Knowledge Discovery and Data Mining*, volume 7301 of *Lecture Notes in Computer Science*, pages 122–134. Springer, Berlin, Heidelberg, 2012.

[114] John H. Holland. *Adaptation in Natural and Artificial Systems: An Introductory Analysis with Applications to Biology, Control, and Artificial Intelligence.* Ann Arbor, MI: University of Michigan Press, 1975.

[115] John H. Holland and Judith S. Reitman. Cognitive systems based on adaptive algorithms. *SIGART Bull.*, 1(63):49–49, June 1977.

[116] Hsiao-Ping Hsu, Vishal Mehra, Walter Nadler, and Peter Grassberger. Growth algorithms for lattice heteropolymers at low temperatures. *The Journal of Chemical Physics*, 118(1):444–451, 2003.

[117] Jingcao Hu and Radu Marculescu. Energy-aware mapping for tile based noc architectures under performance constraints. In *ASPDAC: Proceedings of the 2003 Conference on Asia South Pacific Design Automation*, pages 233–239. ACM, 2003.

[118] Jin Huang and C.X. Ling. Using AUC and accuracy in evaluating learning algorithms. *Knowledge and Data Engineering, IEEE Transactions on*, 17(3):299–310, March 2005.

[119] Richard S.P. Huang, Elena Nedelcu, Yu Bai, Amer Wahed, Kimberly Klein, Hlaing Tint, Igor Gregoric, Manish Patel, Biswajit Kar, Pranav Loyalka, et al. Post-operative bleeding risk stratification in cardiac pulmonary bypass patients using artificial neural network. *Annals of Clinical & Laboratory Science*, 45(2):181–186, 2015.

[120] Yueh-Min Huang, Chun-Min Hung, and Hewijin Christine Jiau. Evaluation of neural networks and data mining methods on a credit assessment task for class imbalance problem. *Nonlinear Analysis: Real World Applications*, 7(4):720–747, 2006.

[121] Simon Huband, Philip Hingston, Luigi Barone, and R. Lyndon While. A review of multi-objective test problems and a scalable test problem toolkit. *IEEE Trans. Evolutionary Computation*, 10(5):477–506, 2006.

[122] Christian Igel, Nikolaus Hansen, and Stefan Roth. Covariance matrix adaptation for multi-objective optimization. *Evolutionary Computation*, 15(1):1–28, 2007.

[123] Hisao Ishibuchi, Noritaka Tsukamoto, and Yusuke Nojima. Evolutionary many-objective optimization. In *Genetic and Evolving Systems, 2008. GEFS 2008. 3rd International Workshop on*, pages 47–52. IEEE, 2008.

[124] J. L. Jacob J. A. Clark, and S. Stepney. The design of S-boxes by simulated annealing. *New Generation Computing*, 23(3):219–231, 2005.

[125] R. G. Jacquot. *Modern Digital Control Systems 2e.* Electrical and Computer Engineering Series. Marcel Dekker, 1995.

[126] Anil K. Jain and Richard C. Dubes. *Algorithms for Clustering Data.* Prentice Hall, Inc., Upper Saddle River, NJ, USA, 1988.

[127] Rabindra Ku Jena and Gopal Ku. Sharma. A multi-objective evolutionary algorithm based optimization model for network-on-chip synthesis. In *Proceedings of ITNG*, pages 977–982. IEEE, 2007.

[128] Dongli Jia, Guoxin Zheng, and Muhammad Khurram Khan. An effective memetic differential evolution algorithm based on chaotic local search. *Information Sciences*, 181(15):3175–3187, 2011.

[129] Taeho Jo and Nathalie Japkowicz. Class imbalances versus small disjuncts. *SIGKDD Explor. Newsl.*, 6(1):40–49, June 2004.

[130] O. Johansson, D. Pearce, and D. Maddison. *Blueprint 5: True Costs of Road Transport*, volume 5. Routledge, 2014.

[131] Nicolas Jozefowiez, Frédéric Semet, and El-Ghazali Talbi. Multi-objective vehicle routing problems. *European Journal of Operational Research*, 189(2):293–309, 2008.

[132] Walsh K. G. Beauchamp. *Functions and Their Applications.* New York: Academic, 1975.

[133] H. Kakigano, Y. Miura, and T. Ise. Distribution voltage control for dc microgrids using fuzzy control and gain-scheduling technique. *Power Electronics, IEEE Transactions on*, 28(5):2246–2258, 2013.

[134] Y. Kanthaphayao and V. Chunkag. Current-sharing bus and fuzzy gain scheduling of proportional-integral controller to control a parallel-connected ac/dc converter. *Power Electronics, IET*, 7(10):2525–2532, 2014.

[135] J. H. Kiefer. Effects of moisture contamination and welding parameters on diffusible hydrogen. *Welding Journal-Including Welding Research Supplement*, 75(5):155–161, 1996.

[136] Il Yong Kim and O. L. De Weck. Adaptive weighted sum method for multi-objective optimization: A new method for pareto front generation. *Structural and Multidisciplinary Optimization*, 31(2):105–116, 2006.

[137] J. Knowles, L. Thiele, and E. Zitzler. A tutorial on the performance assessment of stochastic multi-objective optimizers. 214, Computer Engineering and Networks Laboratory (TIK), ETH Zurich, Switzerland, February 2006. Revised version.

[138] D. J. Kotecki and R. A. LaFave. Aws a5 committee studies of weld metal diffusible hydrogen. *Welding Journal*, 64(3):31–37, 1985.

[139] Sindo Kou. *Welding metallurgy*. John Wiley and Sons, 2003.

[140] Natalio Krasnogor, B. P. Blackburne, Edmund K. Burke, and Jonathan D. Hirst. Multimeme algorithms for protein structure prediction. In *Parallel problem Solving from Nature—PPSN VII*, pages 769–778. Springer, 2002.

[141] Elmar Krieger, Keehyoung Joo, Jinwoo Lee, Jooyoung Lee, Srivatsan Raman, James Thompson, Mike Tyka, David Baker, and Kevin Karplus. Improving physical realism, stereochemistry, and side-chain accuracy in homology modeling: four approaches that performed well in CASP8. *Proteins: Structure, Function, and Bioinformatics*, 77(S9):114–122, 2009.

[142] M. Krzaczek and Z. Kowalczuk. Gain scheduling control applied to thermal barrier in systems of indirect passive heating and cooling of buildings. *Control Engineering Practice*, 20(12):1325–1336, 2012.

[143] M. Kubat and S. Matwin. Addressing the curse of imbalanced training sets: One-sided selection. In *14th International Conference on Machine Learning(ICML97)*, pages 179–186, 1997.

[144] M. Kwan. Reducing the gate count of bitslice DES, 2000.

[145] Kit Fun Lau and Ken A. Dill. A lattice statistical mechanics model of the conformational and sequence spaces of proteins. *Macromolecules*, 22(10):3986–3997, 1989.

[146] J. Laurikkala. Improving identification of difficult small classes by balancing class distribution. In *8th Conference on AI in Medicine in Europe(AIME01)*, volume 2001 of *Lecture Notes on Computer Science*, pages 63–66. Springer Berlin / Heidelberg, 2001.

[147] Tang Lei and Shashi Kumar. A two-step genetic algorithm for mapping task graphs to a network on chip architecture. In *Proceedings of the International Conference on Digital Systems Design*, pages 180–189. IEEE, 2003.

[148] Neal Lesh, Michael Mitzenmacher, and Sue Whitesides. A complete and effective move set for simplified protein folding. In *Proceedings of the Seventh Annual International Conference on Research in Computational Molecular Biology*, pages 188–195. ACM, 2003.

[149] Hui Li and Qingfu Zhang. Multi-objective optimization problems with complicated Pareto sets, MOEA/D and NSGA-II. *IEEE J. EVC*, 13(2):284–302, 2009.

[150] K. Li, S. Kwong, Q. Zhang, and K. Deb. Interrelationship-based selection for decomposition multi-objective optimization. *Cybernetics, IEEE Transactions on*, PP(99):1–1, 2014.

[151] Zhang Liang, Xia Chunming, Cao Jiabin, and Zheng Jianrong. Physical-based modeling of nonlinearities in process control valves. In *Control Engineering and Communication Technology (ICCECT), 2012 International Conference on*, pages 75–78, Dec 2012.

[152] Cheng-Jian Lin and Shih-Chieh Su. Protein 3 d hp model folding simulation using a hybrid of genetic algorithm and particle swarm optimization. *International Journal of Fuzzy Systems*, 13(2):140–147, 2011.

[153] Hai Lin Liu, Fangqing Gu, and Qingfu Zhang. Decomposition of a multi-objective optimization problem into a number of simple multi-objective subproblems. *Evolutionary Computation, IEEE Transactions on*, 18(3):450–455, June 2014.

[154] Y. Liu, J. Ma, J. Niu, Y. Zhang, and W. Wang. Roadside units deployment for content downloading in vehicular networks. In *IEEE International Conference on Communications*, pages 6365–6370, June 2013.

[155] Yi-Hung Liu and Yen-Ting Chen. Face recognition using total margin-based adaptive fuzzy support vector machines. *IEEE Transactions on Neural Networks*, 18(1):178–192, 2007.

[156] C. Lochert, B. Scheuermann, C. Wewetzer, A. Luebke, and M. Mauve. Data aggregation and roadside unit placement for a VANET traffic information system. In *5th ACM Int. Workshop on Vehicular Inter-Networking*, pages 58–65, New York, NY, USA, 2008. ACM.

[157] Rushi Longadge, Snehalata Dongre, and Latesh Malik. Class imbalance problem in data mining review. *International Journal of Computer Science and Network*, 2(1), 2013.

[158] V. López, A. Fernández, and F. Herrera. A first approach for cost-sensitive classification with linguistic genetic fuzzy systems in imbalanced data-sets. In *Intelligent Systems Design and Applications (ISDA), 2010 10th International Conference on*, pages 676–681, Nov 2010.

[159] V. Lopez, A. Fernandez, and F. Herrera. Addressing covariate shift for genetic fuzzy systems classifiers: A case of study with farc-hd for imbalanced datasets. In *Fuzzy Systems (FUZZ), 2013 IEEE International Conference on*, pages 1–8, July 2013.

[160] Victoria López, Sara del Río, José Manuel Benítez, and Francisco Herrera. Cost-sensitive linguistic fuzzy rule based classification systems under the mapreduce framework for imbalanced big data. *Fuzzy Sets and Systems*, 258:5–38, 2015. Special issue: Uncertainty in Learning from Big Data.

[161] Victoria López, Alberto Fernández, María José del Jesus, and Francisco Herrera. A hierarchical genetic fuzzy system based on genetic programming for addressing classification with highly imbalanced and borderline data-sets. *Knowledge-Based Systems*, 38(0):85–104, 2013. Special Issue on Advances in Fuzzy Knowledge Systems: Theory and Application.

[162] Victoria Lopez, Alberto Fernandez, Jose G. Moreno-Torres, and Francisco Herrera. Analysis of preprocessing vs. cost-sensitive learning for imbalanced classification. Open problems on intrinsic data characteristics. *Expert Syst. Appl.*, 39(7):6585–6608, 2012.

[163] I. G. Machado. *Soldagem e Técnicas Conexas: Processos.* Editado pelo Autor, Distribuído pela Associação Brasileira de Soldagem, Porto Alegre, Janeiro, 1996.

[164] M. Mahdizadeh and M. Eftekhari. Designing fuzzy imbalanced classifier based on the subtractive clustering and genetic programming. In *Fuzzy Systems (IFSC), 2013 13th Iranian Conference on*, pages 1–6, Aug 2013.

[165] Thomas W. Malone, Robert Laubacher, and Chrysanthos Dellarocas. Harnessing crowds: Mapping the genome of collective intelligence. 2009.

[166] H. B. Mann and D. R. Whitney. On a test of whether one of two random variables is stochastically larger than the other. *Ann. Math. Statist.*, 18(1):50–60, 03 1947.

[167] R. Timothy Marler and Jasbir S. Arora. The weighted sum method for multi-objective optimization: New insights. *Structural and Multidisciplinary Optimization*, 41(6):853–862, 2010.

[168] R. Massobrio, S. Bertinat, J. Toutouh, S. Nesmachnow, and E. Alba. Smart placement of RSU for vehicular networks using multi-objective evolutionary algorithms. In *2nd Latin America Congress on Computational Intelligence*, pages 1–6, Curitiba, Brazil, October 2015.

[169] R. Massobrio, G. Fagúndez, and S. Nesmachnow. Planificación multiobjetivo de viajes compartidos en taxis utilizando un micro algoritmo evolutivo paralelo. In *X Congreso Español de Metaheurísticas, Algoritmos Evolutivos y Bioinspirados*, 2015. Text in Spanish.

[170] R. Massobrio, J. Toutouh, and S. Nesmachnow. A multi-objective evolutionary algorithm for infrastructure location in vehicular networks. In *7th European Symposium on Computational Intelligence and Mathematics*, pages 1–6, 2015.

[171] M. Matsui. Linear cryptanalysis method for des cipher. In *Advances in Cryptology*, T. Helleseth (Ed.), *LNCS 765*, pages 386–397. Springer-Verlag, 1994.

[172] Warren Lee McCabe, Julian Cleveland Smith, and Peter Harriott. *Unit Operations of Chemical Engineering*, volume 5. McGraw-Hill, New York, 1993.

[173] P. E. McKnight and J. Najab. Kruskal-Wallis test. *Corsini Encyclopedia of Psychology*, 2010.

[174] S. Mendes, J. Gómez-Pulido, M. Vega-Rodríguez, J. Sánchez-Pérez, Y. Sáez, and P. Isasi. The radio network design optimization problem. In *Biologically-Inspired Optimisation Methods*, pages 219–260. Springer Science + Business Media, 2009.

[175] A. J. Menezes, P. C. Van Oorschot, and S. A. Vanstone. *Handbook of Applied Cryptography*. CRC Press, 1996.

[176] K. Miettinen. *Nonlinear Multi-Objective Optimization*. Kluwer, Norwell, MA, 1999.

[177] W. Millan, L. Burnett, G. Cater, J. A. Clark, and E. Dawson. Evolutionary heuristics for finding cryptographically strong S-boxes. In *LNCS 1726*, pages 263–274. Springer-Verlag, 1999.

[178] J. Monteiro, D. Devadas, A. Gosh, K. Keutzer, and J. White. Estimation of average switching activity in combinational logic circuits using symbolic simulation. *IEEE Transactions on Computer-Aided Design of Integrated Circuits and Systems*, 16(1):121–127, 1997.

[179] Douglas C. Montgomery. *Design and Analysis of Experiments*. John Wiley & Sons, 2008.

[180] Área de Movilidad, Ayuntamiento de Málaga. [online], available in `http://movilidad.malaga.eu/`. Retrieved December 2015.

[181] Dinara Mukhlisullina, Andrea Passerini, and Roberto Battiti. Learning to diversify in complex interactive multi-objective optimization. In *Metaheuristics International Conference (MIC 2013)*, 2013.

[182] Srinivasan Murali and Giovanni De Micheli. Bandwidth-constrained mapping of cores onto noc architectures. In *Proceedings of DATE*, pages 896–903. IEEE, 2004.

[183] J. F. Nash. Equilibrium points in n-person games. In *Proceedings of the National Academy of Sciences*, volume 36, pages 48–49, 1950.

[184] J. N. Nash. Non-cooperative game. *Annals of Mathematics*, 54(2):286–295, 1951.

[185] Nadia Nedjah, Marcus Vinícius Carvalho da Silva, and Luiza de Macedo Mourelle. Preference-based multi-objective evolutionary algorithms for power-aware application mapping on noc platforms. *Expert System Applications*, 39(3):2771–2782, 2012.

[186] Nadia Nedjah and Luiza de Macedo Mourelle. Evolutionary multi-objective optimisation: A survey. *International Journal of Bio-Inspired Computation*, 7(1):1–25, 2015.

[187] S. Nesmachnow. An overview of metaheuristics: Accurate and efficient methods for optimisation. *International Journal of Metaheuristics*, 3(4):320–347, 2014.

[188] S. Nesmachnow and S. Iturriaga. Multi-objective scheduling on distributed heterogeneous computing and grid environments using a parallel micro-CHC evolutionary algorithm. In *IEEE International Conference on P2P, Parallel, Grid, Cloud and Internet Computing*, 2011.

[189] S. Nesmachnow and S. Iturriaga. Multi-objective grid scheduling using a domain decomposition based parallel micro evolutionary algorithm. *International Journal of Grid and Utility Computing*, 4(1):70, 2013.

[190] University of Southern California. The network simulator ns-2. [online], available in `http://www.isi.edu/nsnam/ns`. Retrieved December 2015.

[191] National Institute of Standard and Technology. Data encryption standard, federal information processing standards, November 1977.

[192] Umit Y. Ogras, Jingcao Hu, and Radu Marculescu. Key research problems in noc design: A holistic perspective. In *CODES+ISSS: Proceedings of the 3rd IEEE/ACM/IFIP International Conference on Hardware/Software Codesign and System Synthesis*, pages 69–74. ACM, 2005.

[193] Marco Aurélio Cavalcanti Pacheco. Algoritmos genéticos: Princípios e aplicações. *ICA: Laboratório de Inteligência Computacional Aplicada. Departamento de Engenharia Elétrica. Pontifícia Universidade Católica do Rio de Janeiro. Fonte desconhecida*, 1999.

[194] A.M. Palacios, L. Sanchez, and I. Couso. Preprocessing vague imbalanced datasets and its use in genetic fuzzy classifiers. In *Fuzzy Systems (FUZZ), 2010 IEEE International Conference on*, pages 1–8, July 2010.

[195] Ana Palacios, Krzysztof Trawiński, Oscar Cordón, and Luciano Sánchez. Cost-sensitive learning of fuzzy rules for imbalanced classification problems using furia. *International Journal of Uncertainty, Fuzziness and Knowledge-Based Systems*, 22(05):643–675, 2014.

[196] P. Patil and A. Gokhale. Voronoi-based placement of road-side units to improve dynamic resource management in VANETs. In *Int. Conf. on Collaboration Technologies and Systems*, pages 389–396, May 2013.

[197] Dan Pelleg, Andrew W. Moore, et al. X-means: Extending k-means with efficient estimation of the number of clusters. In *ICML*, pages 727–734, 2000.

[198] Jella Pfeiffer, Uli Golle, and Franz Rothlauf. Reference point based multi-objective evolutionary algorithms for group decisions. In *Proceedings of the 10th Annual Conference on Genetic and Evolutionary Computation*, pages 697–704. ACM, 2008.

[199] M. Picone, S. Busanelli, M. Amoretti, F. Zanichelli, and G. Ferrari. *Advanced Technologies for Intelligent Transportation Systems*. Springer International Publishing, 2015.

[200] Adam P. Piotrowski. Adaptive memetic differential evolution with global and local neighborhood-based mutation operators. *Information Sciences*, 241:164–194, 2013.

[201] I. Poikolainen and F. Neri. Differential evolution with concurrent fitness based local search. In *Evolutionary Computation (CEC), 2013 IEEE Congress on*, pages 384–391, June 2013.

[202] A. Ponsich, A.L. Jaimes, and C.A.C. Coello. A survey on multi-objective evolutionary algorithms for the solution of the portfolio optimization problem and other finance and economics applications. *Evolutionary Computation, IEEE Transactions on*, 17(3):321–344, June 2013.

[203] Ronaldo C. Prati, Gustavo E.A.P.A. Batista, and Maria Carolina Monard. Class imbalances versus class overlapping: An analysis of a learning system behavior. In Raúl Monroy, Gustavo Arroyo-Figueroa, Luis Enrique Sucar, and Humberto Sossa, editors, *MICAI 2004: Advances in Artificial Intelligence*, volume 2972 of *Lecture Notes in Computer Science*, pages 312–321. Springer, Berlin, Heidelberg, 2004.

[204] Bin Qian, Angel R. Ortiz, and David Baker. Improvement of comparative model accuracy by free-energy optimization along principal components of natural structural variation. *Proceedings of the National*

Academy of Sciences of the United States of America, 101(43):15346–15351, 2004.

[205] Ciprian Radu. *Optimized Algorithms for Network-on-Chip Application Mapping ANoC*. PhD thesis, University of Sibiu, Romenia, 2011.

[206] M. D. Ciprian Radu, Shahriar Mahbub, and Lucian Vintan. Developing domain-knowledge evolutionary algorithms for network-on-chip application mapping. *Microprocessors and Microsystems - Embedded Hardware Design*, 37(1):65–78, 2013.

[207] Srivatsan Raman, Yuanpeng J. Huang, Binchen Mao, Paolo Rossi, James M. Aramini, Gaohua Liu, Gaetano T. Montelione, and David Baker. Accurate automated protein NMR structure determination using unassigned NOESY data. *Journal of the American Chemical Society*, 132(1):202–207, 2009.

[208] D. Ramyachitra and P. Manikandan. Imbalanced dataset classification and solutions: A review. *International Journal of Computing and Business Research*, 5(4), 2014.

[209] A. Reis, S. Sargento, F. Neves, and O. Tonguz. Deploying roadside units in sparse vehicular networks: What really works and what does not. *IEEE Transactions on Vehicular Technology*, 63(6):2794–2806, 2014.

[210] Margarita Reyes-Sierra and Carlos A. Coello Coello. Multi-objective particle swarm optimizers: A survey of the state-of-the-art. *International Journal of Computational Intelligence Research*, 2(3):287–308, 2006.

[211] G. Reynoso-Meza, J. Sanchis, X. Blasco, and J.M. Herrero. Hybrid DE algorithm with adaptive crossover operator for solving real-world numerical optimization problems. In *Evolutionary Computation (CEC), 2011 IEEE Congress on*, pages 1551–1556, June 2011.

[212] V. T. Rhyne. *Fundamentals of Digital Systems Design*. Prentice Hall Electrical Engineering Series, 1973.

[213] Daniela Röthlisberger, Olga Khersonsky, Andrew M. Wollacott, Lin Jiang, Jason DeChancie, Jamie Betker, Jasmine L. Gallaher, Eric A. Althoff, Alexandre Zanghellini, Orly Dym, et al. Kemp elimination catalysts by computational enzyme design. *Nature*, 453(7192):190–195, 2008.

[214] Roberto Santana, Pedro Larrañaga, and José A. Lozano. Protein folding in 2-dimensional lattices with estimation of distribution algorithms. In *Biological and Medical Data Analysis*, pages 388–398. Springer, 2004.

[215] Roberto Santana, Pedro Larrañaga, and Jose A. Lozano. Protein folding in simplified models with estimation of distribution algorithms. *Evolutionary Computation, IEEE Transactions on*, 12(4):418–438, 2008.

[216] Hiroyuki Sato. Inverted PBI in MOEA/D and its impact on the search performance on multi and many-objective optimization. In *Proceedings of the 2014 Conference on Genetic and Evolutionary Computation*, GECCO '14, pages 645–652, New York, NY, USA, 2014. ACM.

[217] Hiroyuki Sato. Analysis of inverted pbi and comparison with other scalarizing functions in decomposition based MOEAS. *Journal of Heuristics*, 21(6):819–849, 2015.

[218] S. Saunders and A. Aragon. *Antennae and Propagation for Wireless Communication Systems*. Wiley, New York, NY, USA, 1999.

[219] O. Schutze, X. Esquivel, A. Lara, and Carlos A. Coello Coello. Using the averaged Hausdorff distance as a performance measure in evolutionary multi-objective optimization. *Evolutionary Computation, IEEE Transactions on*, 16(4):504–522, Aug 2012.

[220] C. E. Shanon. Communication theory of secrecy systems. 28(4):656–715, 1949.

[221] Yang Shen, Robert Vernon, David Baker, and Ad Bax. De novo protein structure generation from incomplete chemical shift assignments. *Journal of Biomolecular NMR*, 43(2):63–78, 2009.

[222] Alena Shmygelska, Rosalia Aguirre-Hernandez, and Holger H. Hoos. An ant colony optimization algorithm for the 2d HP protein folding problem. In *Ant Algorithms*, pages 40–52. Springer, 2002.

[223] Alena Shmygelska and Holger H. Hoos. An improved ant colony optimisation algorithm for the 2d hp protein folding problem. In *Advances in Artificial Intelligence*, pages 400–417. Springer, 2003.

[224] Stephen Frederick Smith. *A Learning System Based on Genetic Adaptive Algorithms*. PhD thesis, University of Pittsburgh, Pittsburgh, PA, USA, 1980. AAI8112638.

[225] Rainer Storn. On the usage of differential evolution for function optimization. In *NAFIPS'96*, pages 519–523. IEEE, 1996.

[226] Rainer Storn and Kenneth Price. Differential evolution – A simple and efficient heuristic for global optimization over continuous spaces. *J Glob. Opt.*, 11(4):341–359, 1997.

[227] Xiaoyan Sun, Lei Yang, Dunwei Gong, and Ming Li. Interactive genetic algorithm assisted with collective intelligence from group decision making. In *Evolutionary Computation (CEC), 2012 IEEE Congress on*, pages 1–8. IEEE, 2012.

[228] R. K. Sundaram. A first course in optimization theory. Cambridge University, Press, Cambridge, 1996.

[229] James Surowiecki. *The Wisdom of Crowds.* Random House LLC, 2005.

[230] T. Takagi and M. Sugeno. Fuzzy identification of systems and its applications to modeling and control. *Systems, Man and Cybernetics, IEEE Transactions on*, SMC-15(1):116–132, Jan 1985.

[231] Ryoji Tanabe and Alex Fukunaga. *Parallel Problem Solving from Nature – PPSN XIII: 13th International Conference, Ljubljana, Slovenia, September 13-17, 2014. Proceedings*, chapter Reevaluating Exponential Crossover in Differential Evolution, pages 201–210. Springer International Publishing, Cham, 2014.

[232] Lothar Thiele, Kaisa Miettinen, Pekka J. Korhonen, and Julian Molina. A preference-based evolutionary algorithm for multi-objective optimization. *Evolutionary Computation*, 17(3):411–436, 2009.

[233] I. Tomek. Two modifications of cnn. *IEEE Transactions on Systems and Man and Cybernetics*, 6:769–772, 1976.

[234] O. Trullols, M. Fiore, C. Casetti, C. Chiasserini, and J. Ordinas. Planning roadside infrastructure for information dissemination in intelligent transportation systems. *Computer Communications*, 33(4):432–442, 2010.

[235] Che-Hui Tsai, Li-Chiu Chang, and Hsu-Cherng Chiang. Forecasting of ozone episode days by cost-sensitive neural network methods. *Sci. Total Environ*, 407(6):2124–35, 2009.

[236] Ron Unger and John Moult. Genetic algorithms for protein folding simulations. *Journal of Molecular Biology*, 231(1):75–81, 1993.

[237] David A Van Veldhuizen and Gary B. Lamont. Evolutionary computation and convergence to a Pareto front. In *Late Breaking Papers at the Genetic Programming 1998 Conference*, pages 221–228. Citeseer, 1998.

[238] Pedro Villar, Alberto Fernández, Ramón A. Carrasco, And Francisco Herrera. Feature selection and granularity learning in genetic fuzzy rule-based classification systems for highly imbalanced data-sets. *International Journal of Uncertainty, Fuzziness and Knowledge-Based Systems*, 20(03):369–397, 2012.

[239] H.O. Wang, K. Tanaka, and M. Griffin. Parallel distributed compensation of nonlinear systems by Takagi–Sugeno fuzzy model. In *Fuzzy Systems, 1995. International Joint Conference of the Fourth IEEE International Conference on Fuzzy Systems and the Second International Fuzzy Engineering Symposium, Proceedings of 1995 IEEE Int*, volume 2, pages 531–538, Mar 1995.

[240] Ling Wang, Eric A. Althoff, Jill Bolduc, Lin Jiang, James Moody, Jonathan K Lassila, Lars Giger, Donald Hilvert, Barry Stoddard, and David Baker. Structural analyzes of covalent enzyme–substrate analog complexes reveal strengths and limitations of de novo enzyme design. *Journal of Molecular Biology*, 415(3):615–625, 2012.

[241] Andrzej P. Wierzbicki. The use of reference objectives in multi-objective optimization. In *Multiple Criteria Decision Making Theory and Application*, pages 468–486. Springer, 1980.

[242] Frank Wilcoxon. Individual comparisons by ranking methods. *Biometrics Bulletin*, 1(6):80–83, December 1945.

[243] D. P. Williams, V. Myers, and M. S. Silvious. Mine classification with imbalanced data. *Geoscience and Remote Sensing Letters, IEEE*, 6(3):528–532+, 2009.

[244] E. Williams. Aviation formulary v1.46. [online], available in `http://williams.best.vwh.net/avform.htm`. Retrieved August 2015.

[245] Dennis L. Wilson. Asymptotic properties of nearest neighbor rules using edited data. *Systems, Man and Cybernetics, IEEE Transactions on*, 2(3):408–421, July 1972.

[246] Y. Xiong, J. Ma, W. Wang, and J. Niu. Optimal roadside gateway deployment for VANETs. *Przeglad ElektroTechniczny*, (7):273–276, 2012.

[247] Hongjiu Yang, Xuan Li, Zhixin Liu, and Ling Zhao. Robust fuzzy-scheduling control for nonlinear systems subject to actuator saturation via delta operator approach. *Information Sciences*, 272(0):158–172, 2014.

[248] L. A. Zadeh. Fuzzy sets. *Information and Control*, (8):338–353, 1965.

[249] Lofti A. Zadeh. Fuzzy sets. *Information and Control*, 8:338–353, 1965.

[250] Saúl Zapotecas-Martínez and Carlos A. Coello Coello. Monss: A multi-objective nonlinear simplex search approach. *Engineering Optimization*, 48(1):16–38, 2016.

[251] Gustavo R. Zavala, Antonio J. Nebro, Francisco Luna, and Carlos A. Coello Coello. A survey of multi-objective metaheuristics applied to structural optimization. *Structural and Multidisciplinary Optimization*, 49(4):537–558, 2014.

[252] Chunmei Zhang, Jie Chen, and Bin Xin. Distributed memetic differential evolution with the synergy of Lamarckian and Baldwinian learning. *Applied Soft Computing*, 13(5):2947–2959, 2013.

[253] L. B. Zhang, C. G. Zhou, X. H. Liu, Z. Q. Ma, M. Ma, and Y. C. Liang. Solving multi-objective optimization problems using particle swarm optimization. In *Proceedings of the Congress on Evolutionary Computation*, pages 2400–2405. IEEE, 2003.

[254] Q. Zhang and H. Li. MOEA/D: A multi-objective evolutionary algorithm based on decomposition. *IEEE J. Evc.*, 11(6):712–731, 2007.

[255] Q. Zhang, A. Zhou, S. Zhao, P. N. Suganthan, W. Liu, and S. Tiwari. Multi-objective optimization test instances for the CEC 2009 special session and competition. *Mechanical Engineering*, pages 1–30, 2009.

[256] Qingfu Zhang, Wudong Liu, and Hui Li. The performance of a new version of MOEA/D on CEC09 unconstrained mop test instances. In *Evolutionary Computation, 2009. CEC '09. IEEE Congress on*, pages 203–208, May 2009.

[257] Wenbiao Zhou, Yan Zhang, and Zhigang Mao. Pareto based multi-objective mapping ip cores onto noc architectures. In *Proceedings of APCCAS*, pages 331–334. IEEE, 2006.

[258] E. Zitzler, M. Laumanns, and L. Thiele. SPEA2: Improving the strength Pareto evolutionary algorithm for multi-objective optimization. In *Evolutionary Methods for Design Optimization and Control with Applications to Industrial Problems*, pages 95–100, Athens, Greece, 2001.

[259] E. Zitzler, M. Laumanns, and L. Thiele. SPEA2: Improving the strength Pareto evolutionary algorithm for multi-objective optimization. In K. C. Giannakoglou, D. T. Tsahalis, J. Périaux, K. D. Papailiou, and T. Fogarty, editors, *Evolutionary Methods for Design Optimization and Control with Applications to Industrial Problems*, pages 95–100, Athens, Greece, 2001. International Center for Numerical Methods in Engineering.

[260] E. Zitzler and L. Thiele. Multi-objective evolutionary algorithms: A comparative case study and the strength Pareto evolutionary algorithm. *IEEE J EVC*, 3(4):257–271, 1999.

[261] Eckart Zitzler, Dimo Brockhoff, and Lothar Thiele. The hypervolume indicator revisited: On the design of Pareto-compliant indicators via weighted integration. In *Evolutionary Multi-criterion Optimization*, pages 862–876. Springer, 2007.

[262] Eckart Zitzler, Kalyanmoy Deb, and Lothar Thiele. Comparison of multi-objective evolutionary algorithms: Empirical results. *Evolutionary Computation*, 8(2):173–195, 2000.

[263] Eckart Zitzler, Kalyanmoy Deb, and Lothar Thiele. Comparison of multi-objective evolutionary algorithms: Empirical results. *Evolutionary Computation*, 8(2):173–195, 2000.

[264] Eckart Zitzler, Kalyanmoy Deb, and Lothar Thiele. Comparison of multi-objective evolutionary algorithms: Empirical results. *Evolutionary Computation*, 8:173–195, 2000.

[265] Eckart Zitzler, Joshua Knowles, and Lothar Thiele. Quality assessment of Pareto set approximations. In Jürgen Branke, Kalyanmoy Deb, Kaisa Miettinen, and Roman Slowinski, editors, *Multi-Objective Optimization: Interactive and Evolutionary Approaches*, volume 5252 of *Lecture Notes in Computer Science*, pages 373–404. Springer, Berlin, Heidelberg, 2008.

[266] Eckart Zitzler and Simon Künzli. Indicator-based selection in multi-objective search. In *Parallel Problem Solving from Nature-PPSN VIII*, pages 832–842. Springer, 2004.

[267] Eckart Zitzler and Simon Künzli. *Parallel Problem Solving from Nature - PPSN VIII: 8th International Conference, Birmingham, UK, September 18-22, 2004. Proceedings*, chapter Indicator-Based Selection in Multi-Objective Search, pages 832–842. Springer, Berlin, Heidelberg, Berlin, Heidelberg, 2004.

[268] Eckart Zitzler, Marco Laumanns, and Lothar Thiele. SPEA2: Improving the strength Pareto evolutionary algorithm, 2001.

[269] Eckart Zitzler and Lothar Thiele. Multi-objective optimization using evolutionary algorithms – A comparative case study. In *Parallel Problem Solving from Nature—PPSN V*, pages 292–301. Springer, 1998.

[270] Eckart Zitzler, Lothar Thiele, Marco Laumanns, Carlos M. Fonseca, and Viviane Grunert Da Fonseca. Performance assessment of multi-objective optimizers: An analysis and review. *Evolutionary Computation, IEEE Transactions on*, 7(2):117–132, 2003.

Index